"*The Immoral Majority* is a fascinating look inside a world that's a mystery to non-Evangelicals like me. But it's also something I didn't expect: funny and moving. This book will be read for years as an explanation not only of the Trump era, but of the political rise and spiritual fall of an entire movement in twenty-first-century America."

—Tom Nichols, author of *The Death of Expertise*

"Ben Howe has crafted a book that lays out the rocky path the bulk of Evangelical Christian movement took, going from being the arbiters of morality, to shrugging it off and embracing Donald Trump in service to power."

—Jay Caruso, deputy editor, *Washington Examiner*

THE IMMORAL MAJORITY

The

IMMORAL

MAJORITY

Why Evangelicals Chose
Political Power over Christian Values

BEN HOWE

BROADSIDE BOOKS
An Imprint of HarperCollinsPublishers

FIRST BROADSIDE PAPERBACK EDITION PUBLISHED 2020.

The Library of Congress has catalogued the hard cover edition as follows:

Names: Howe, Ben (Conservative blogger), author.
Title: The immoral majority : why evangelicals chose political power over
   Christian values / Ben Howe.
Description: First edition. | New York : Broadside, 2019. | Includes index.
Identifiers: LCCN 2019022008 | ISBN 9780062797117 (hardcover)
Subjects: LCSH: Christianity and politics—United States. | Christian
   ethics—United States. | Evangelicalism—United States. | Conduct of
   life. | United States—Church history—21st century.
Classification: LCC BR516 .H67 2019 | DDC 261.0973—dc23
LC record available at https://lccn.loc.gov/2019022008

ISBN 978-0-06279708-7 (pbk.)

20 21 22 23 24   LSC   10 9 8 7 6 5 4 3 2 1

For Mia, Abby, Colin, and Chloe. Everything is for you.
For Mom and Dad. Everything is from you.

CONTENTS

INTRODUCTION

On August 1, 2012, my friend Bruce picked me up and we headed to Chick-fil-A for lunch. We were in Fort Mill, South Carolina, and I had my camera in hand to ask Bruce about why this was important to him.

At the time Chick-fil-A was in the news everywhere after well over a year of a growing controversy regarding millions of dollars in donations the fast-food company had made via its charitable foundation to what the *Huffington Post* described as "groups that have anti-gay agendas," such as the Marriage & Family Legacy Fund, Focus on the Family, and the Family Research Council.[1]

While CEO Dan Cathy had spent about nineteen months pushing back on the implication that his company favored discrimination or was homophobic, the inquisition into the company's ideology as well as the calls by the LGBTQ community to boycott their restaurants was injected with newfound outrage following comments Cathy made in an interview with the Baptist Press on July 16, 2012.

"We are very much supportive of the family, the biblical definition of the family unit," Cathy said. "We are a family-owned business, a family-led business, and we are married to our first wives. We give God thanks for that."

In the supercharged summer of a presidential election year, it's easy to see how a story like this exploded beyond anyone's genuine feelings on the matter, but I and my friend Bruce, himself a conservative blogger under the name Gay Patriot, had found the attacks on the company dishonest.

While there could be reasonable debates over the organizations that had received donations, or even just basic disagreement with the perspective Cathy offered on marriage, we felt Chick-fil-A was well known for being a welcoming restaurant with great food and excellent service.

It seemed to us that while the company's owners may hold positions that reflected their religious values, there was no evidence that those positions would trickle down in a negative way toward gay customers and, given that Bruce himself is gay, we sought to prove that point on video.

So, as I said, we traveled to the Chick-fil-A.

"I'm gonna prove a point that Chick-fil-A is open, and their food is good, and they accept all Americans," he said.

"And you yourself are a gay American?" I asked.

"I am a gay American ... that eats Chick-fil-A," he replied.

When we arrived, the restaurant was overwhelmed with cars in the drive-through. Vehicles had circled the building twice over in apparent support of what had been dubbed "Chick-fil-A Appreciation Day," a "buycott" that seemed to start with a Facebook post by evangelical and former presidential candidate Governor Mike Huckabee.

We parked and walked toward the doors after marveling at the crowd size. Bruce was walking just a bit ahead of me, and I was still filming.

That's when the unthinkable happened.

In the video we put up on YouTube, Bruce could not get near the doors as the result of what appeared to be an invisible barrier of some sort.

We were both very confused, especially after I walked right past where the apparent force field was and found nothing obstructing my entrance.

Before we could alert management or try to determine why this was happening, we heard several loud rumbling sounds, followed by the emergence of something from above the restaurant roof that neither of us could put into words. All we could offer was fear and confusion as I finally yelled to Bruce, "Dude, get out of here!"

Before we knew it, we were bombarded from above with machine-gun fire as well as full-fledged laser beams. I continued to film as we ran for our lives with plasma and bullets flying all around us.

Titled *Not Without My Chicken*, the video of our adventure showcased Bruce's and my Oscar-worthy performances, with the help of some special effects at the direction of my friend Chris Loesch.[2]

Through this video, we wanted to highlight the absurdity of the notion that Chick-fil-A discriminated against gay people like Bruce. It was lighthearted and good-natured, and in the years since its minor viral fame, the video has been praised even by those who won't step foot in the stores.

But despite this bit of fun, there were many moving pieces that week related to the story of Chick-fil-A. The same day we filmed our adventure, I found myself in another one of those stories, a darker one.

It seems that another politically minded individual had decided to go to a Chick-fil-A on the same day with his camera in hand to make the opposite point from the one Bruce and I were trying to make.

Thirty-seven-year-old Arizona resident Adam Smith waited patiently in the drive-through line for a free water so he could film his planned interaction with whoever was working the window at that time.

The target of the video ended up being employee Rachel Elizabeth, who quickly found herself being filmed and having her character questioned for choosing to work at Chick-fil-A.

"You know why I'm getting the free water, right?" Smith asked her.

"I do not," she replied with a smile.

"Because Chick-fil-A is a hateful corporation," he declared, which quickly caused Elizabeth, who could see she was being filmed, to lose her smile.

"I disagree," she said, adding, "we don't treat any of our customers differently; we don't discriminate in hiring practices."

Smith interrupted her and cited the company's donations as hateful, prompting Elizabeth, whose polite smile had returned, to attempt to defuse the disagreement by telling Smith that she was staying neutral on the subject and that she didn't think her personal beliefs belonged in the workplace.

After some more politeness between the two of them, despite the volatile subject, Elizabeth said she was uncomfortable with the fact that he was filming her. "I totally understand," Smith replied.

Altogether it wasn't shaping up to be a very interesting interaction—Smith had said what he thought, and Elizabeth had politely disagreed. That could have been the whole story, and one that probably didn't have much viral potential.

Then things got uncomfortably and inappropriately personal.

After at last handing Smith the water, Elizabeth said, "It's my pleasure to serve you, always," speaking over him as he calmly relished in the fact that he had "taken some money" from Chick-fil-A so that less could be given to "hate groups."

Elizabeth maintained her customer service temperament, voicing additional assurances that she was happy to serve him, but she was interrupted again by Smith, who said, "I don't know how you live with yourself and work here. I don't understand it. This is a horrible corporation with horrible values."

Despite this, she continued to thank him, smile, and tell him it was a pleasure, prompting Smith to add, as he was pulling away, "I'm a nice guy, by the way, and I'm totally heterosexual. Just can't stand the hate. It's gotta stop, you guys, stand up."

Evidently believing that he had done a good deed, Smith uploaded the video upon returning home, seemingly unaware of how bad he looked in it and how undeserving of his treatment Elizabeth appeared to be.

The response would be eye-opening for him.

Given that this was several years ago, I can't recall exactly the moment I saw the video, which didn't stay online long, but I was already pretty active on the social media site Twitter, which at least tells some of the story.

My earliest tweet on the matter was a link to his YouTube video, where I artfully said, "Douchebaggery, thy name is this guy."[3]

"That makes me shake with anger, we need to find out who she is and help her out," one tweeter replied.

"Yep," said another. "Seems like a real tolerant kind of guy. He has free speech and if someone disagrees they are hate groups."

I recall agreeing with both of these sentiments. I was angry, which was helped by how rudely this man, whose identity wasn't known to us, would treat someone who had nothing to do with his narrative and who only smiled and offered kindness to him.

And then I said something that at the time was not as common but as of this writing is a problem invading many aspects of American life.

"We should find out where the guy in this video works and complain to him that his company hires shitbags," I said.[4]

I didn't spearhead the charge, but I was one member of a growing Twitter mob. The hunt began immediately.

By the next day, Smith's video long deleted, Internet sleuths had located his LinkedIn profile and, in short order, his place of work.

I don't know how much I was thinking of Smith as a human in that moment. He was more like a point. A purpose. His existence was irrelevant to me, other than how it could serve what I wanted.

In fairly short order, after his employer, a company called Vante, was bombarded with unwanted media and angry phone calls, Smith was let go from his job as chief financial officer.

Did I feel any sympathy for him? Not really. I tweeted upon the news, "We didn't get this guy fired. He got fired because he embarrassed his company. HE embarrassed his company."[5]

For me and many others, the very people who had used our social media perches to spread this man's identity, the story ended there. Justice had been served.

We were the good guys. They were the bad guys.

We put out a positive video that was fun and sparked conversation. "They" put out a video full of judgment, condemnation, and intolerance. We received accolades and pats on the back. They suffered the consequences of their actions.

While that was a neatly tied-up end to the drama, the story most certainly wasn't over for Adam Smith.

Upon his firing, he took to YouTube again, this time to offer an apology to Rachel Elizabeth for how he'd treated her.

"Rachel, I am so very sorry for the way I spoke to you on Wednesday," he said. "You handled my frustrating rant with such dignity and composure. Every time I watch the video I'm blown away by, really, the beauty in what you did, in your kindness and your patience with me."

He went on to talk about his motivations and his regret, but whether he did this as an attempt to rescue his situation or because he truly felt this way was irrelevant. The damage was done: he was out of a job and he would not be getting it back. A husband and father of four, Smith had had a successful career in the medical industry as the CFO of the aforementioned Vante, a medical supply company. His video stunt cost him his job, following what the company told him had been hundreds of voicemails as well as death threats and bomb threats, and Smith initially assumed that, while his event had been costly, his previously blemish-free career and résumé would help him land on his feet.

Smith managed to find new job in Portland, Oregon, and thought the dust had settled, but two weeks in he found himself in the office of his new employer's CFO.

"You lied to us," they said. "You didn't tell us about this video."

He was fired on the spot.

He soon applied for another job and, after an interview, received an offer. Having learned from the previous experience, he decided to tell them about the video before he accepted the offer. At first it seemed they were okay with it, just asking for an assurance that his days of video activism were over.

But in short order they left him a voicemail informing him that the board had reconsidered and decided to rescind the offer.

Smith said this pattern repeated itself in the years that followed, and his family eventually had to resort to food stamps to survive without his income. He says he became suicidal, contemplating driving off a cliff so his wife and children could collect on his insurance policy, but he couldn't bring himself to do it.

His life had been completely and seemingly irreversibly destroyed.

Stories like this are all too common these days. And, often depending on which side of an ordeal people find themselves, their destruction is often portrayed as either unjust persecution or righteous street justice. At the time that Smith's life was torn apart, it was certainly justice in my mind.

His actions were and still are indefensible. His apology was, at best, questionable. He did it to himself, these were the consequences, and, even as I write this and regret participating, I still am not excluding Smith from the role he played in his own downfall.

It would be easy to wrap up this story with questions about

punishments befitting crimes and offer some thoughts on how there's a point at which justice becomes revenge. And honestly, that takeaway is adequate, even if not enough to be relevant to the purpose of this book.

The real question here isn't whether the destruction of a person's livelihood as a result of his making a three-minute video is a bit harsh. Clearly it is, if just for the sake of his children, who were innocent of his wrongdoing.

Nor is the question whether Smith is ultimately responsible for placing himself in danger. For instance, if a man with four children decided to cheat on his wife, he's certainly responsible for any consequences that may come from that decision. He is the one who decided to act out, and so, regardless of who gets caught in the crossfire of any bad consequences, the onus was on him.

When his wife finds out about the affair, it would be justice for her to divorce him, take him to the cleaners financially, and basically make him feel bad about what he's done. But it would be something else entirely if she chose instead to murder him for his infidelity.

It would be wrong, not only because of the severity of the sentence but because it simply creates more victims, such as their children, who would then have lost their father.

It's not really about whether the punishment fits the crime; it's more about the decisions of those who react to the crime and whether they are carrying out justice or simply joining the wrongdoer in being wrong.

Whatever fallout came from the scenario of adultery above, the cheater deserved what came to him. Absolutely. But that says nothing of the motives of the wife who chose to murder him for it.

He deserved punishment. But it was beyond her moral authority to enact justice to that extreme.

It's the same with Smith. And it's the same with me.

Smith caused the circumstances of his situation and was solely responsible for that. But once I and others chose to participate in doing something about it, we exceeded any mandate we had to do good and chose instead to return evil with evil.

Then we justified what we did by resting on this silly notion of what he deserved, as though we were ever in any position to render such a judgment.

Sympathy is not a factor here. Even reading up on his recovery from this mess and seeing that he has chosen to use it as a way to counsel others out of giving in to the worst parts of themselves, I still doubt I'd even like him that much. I'm not looking to get together with him and have a coffee and reconcile, as though there is some friendship that was lost. I doubt he even noticed I existed among the throngs of people who came after him.

But I was able to remember at some point in the aftermath that he's a human, even if he's one who might grate on me in person. He is someone who has been granted the opportunity for redemption, just as I have.

God doesn't love him less than he loves me. He doesn't want less for his life or the lives of his children than he wants for me and my children.

Social media makes it easy to forget all of that in an instant of impulse. Within a day of watching Smith dehumanize this woman, my immediate reaction instructed me that the only possible response was to dehumanize Smith.

He wasn't a person: he was garbage.

It's been a few years since all of that went down, and since then, and following several events in my own life, I've come to the awareness that I'm not a moral person carrying out justice against immoral people.

The truth is, I'm fully capable of being awful. It's my job to fight that capacity to be awful, not pretend it doesn't exist.

Take a cursory glance at my Twitter feed and you'll get all the evidence you need, with countless examples of my hurling snark and profanity-littered insults at people who have dared to argue with or insult me, people I very clearly believed to be beneath my station.

On second thought, don't go to my Twitter feed.

If we were to meet out in the real world and you expressed interest in who I was and how I came to be this person, you'd hear about more of those pesky moral failings that weave the tapestry of my past. You'd know that while I grew up in a Christian home, I rebelled as a teenager and was involved in drugs and crime. I'm a divorced father of four who has struggled at times with faith in light of the hardships. It doesn't help that I always come around to realizing I was the architect of these hardships.

Though it is possible for me to find myself in a morally superior position to Twitter trolls who are hurling unwarranted insults at me, they are as capable of being better than they are in that moment as I am at being worse.

Although my responsibilities as an adult, and especially as a single father, require me to better tame the demons that can influence my behaviors, I am still capable of being the person I was when I was seventeen. Or thirty-eight. Or any of the other times of my life when I've allowed the worst parts of me to become my most visible traits. Allowing my capacity to be sinful to be unleashed, often with devastating consequences.

The same is true and was true of Adam Smith. And Rachel Elizabeth. And Dan Cathy.

And you.

We believed Smith was garbage and, importantly, we weren't. But I have come to believe that it is vitally important that as human beings we accept that we are *all* capable of being garbage. It has occurred to me that one of the quickest ways to thwart any possibility of personal growth is to assume such a thing is not within you.

The key to being a good person is accepting that capacity to be garbage, as counterintuitive as that may sound.

Because what is it that we've seen happens when people live life assuming they are righteous? What happens when they cannot see that, as an imperfect person with a sinful nature, they are destined to have the capacity for evil?

From what I've seen and at times lived, they will do almost nothing to correct themselves. They hold on to the inaccurate picture they've created. It becomes central to their identity and encourages rationalization and self-deception.

Put thousands of people like this together . . . hundreds of thousands . . . millions? Well, then you've got a right proper problem. An echo chamber is created that has all of the people who subscribe to this notion—that they walk on the side of the angels—feeding off of one another and drunk on confirmation bias.

This is a book about what happens when the people who believe they have the moral high ground find themselves on the low road. Christians in America increasingly face powerful opponents. We are surrounded by atheism and hedonism, yet we keep finding no greater enemy than ourselves. Throughout the following chapters, I present situations where hard choices had to be made. Should we have proceeded by adhering to biblical principles, even if it meant we would keep losing? Would we win more followers defending our values in the public square by leaning into our capacity for brutality, cruelty, and indifference, or by acting according to Christ's example of winsome, compassionate witness?

The dilemmas are familiar to us all, but too often we've bought into the common wisdom about them. Misunderstanding what is happening, we fail to correctly identify what is at stake. We miscalculate the short-term problems while forgetting the long term altogether.

The sources of our greatest strengths—an awareness of the costs of sin and the desire to save others—can be corrupted, becoming instead the tools of our greatest weaknesses. What we need is a new conventional wisdom for Christians living in modern political America. A conventional wisdom that can be realized only if we find the strength to be a light for others, even when we see the light of our nation dimming.

The greatest sin of the Pharisees was their corruption of God's Law. They professed and enforced it, but they did not teach it nor understand it. They used it instead to acquire the world, and in doing so abandoned the very reasons the Law had been created.

Woe to you Pharisees, because you give God a tenth of your mint, rue and all other kinds of garden herbs, but you neglect justice and the love of God. You should have practiced the latter without leaving the former undone.
—LUKE 11:42

This failure to uphold the very standards we claim animate our passions—a weakness not unique to Christians—is the source of all our problems in politics and is the very heart of partisanship.

If you want to quickly gauge yourself on this, ask yourself how you feel reading the Chick-fil-A story I told.

Do you think to yourself, "It sounds as if Smith ended up okay. He suffered, but he's the one who created the situation. I don't see the problem, and I don't see how my sympathy or lack thereof is of any consequence. He needed to learn a lesson, and people taught it to him. Good for them."

Or do you think, "All of those who helped do this to Smith went way over the top, and they all deserve to have the same thing happen to them!"

These are fairly normal reactions, in my experience. Separating the good guys from the bad guys, the right from the wrong. It's often influenced by the starting premise.

Perhaps you saw Smith's actions as tactless, but you agree the larger issue is that Chick-fil-A funds "hate groups."

Or perhaps you believe that Smith represents the larger cultural problems endemic to the progressive left, which you think preaches intolerance and cruelty, and that what we did by going after him sent a message that we're not going to take it anymore.

In the six years between then and now, I have seen this dynamic of seeking justice in service of winning a culture war play out over and over and with increasing intensity and cost.

It is utterly common now for rage mobs to descend on wrongdoers and shame them into oblivion, sometimes finding out later that those accused hadn't even done the alleged wrong.

What makes the toxicity of it all so potent is that, with rare exception, these crusades of revenge are covered in rationalizations of justice. People are being destroyed, either for a wrongdoing or for the perception of one, and many walk away believing they'd done some good in the world. Much like Smith thought when he pulled away from that drive-through window with no idea that his life was about to take a downward turn.

Broadly speaking, we have taken to confronting immorality by becoming immoral. But because our immorality is intended to stop an objectively worse immorality, we reason that it is *not* immoral.

Morality can be subjective of course. Not everyone shares the same values and principles, nor the same beliefs about God or the absence of God.

But when I decided to write this book, it was because I had observed so much of this mentality I displayed in going after Smith and feeling justified by it in the words, actions, and votes of Christians.

It seems as though grace, redemption, compromise, and empathy had all become the dirty words of "losers" who hadn't accepted that larger issues were at stake. I saw more and more the belief that we were amid a culture war and that our job in prosecuting that war was to inflict maximum pain. The conclusion was that pain is the only way our immoral enemies will "learn."

As a human, this is something I can relate to. That scorned wife in my adultery example may have been wrong to murder her cheating husband, but I doubt the jury that convicted her would claim they don't "get it."

I get it, too. When you believe you've suffered, it's natural to want to inflict maximum damage. Of course, we all know that is vengeance, not justice.

So while as a human I get it, as a Christian I can't get behind it.

"Vengeance is Mine; I will repay, says the Lord" (Romans 12:19) is one of those verses that even people who have never picked up a Bible recognize. It's pretty indisputable that seeking vengeance is sinful, so no matter how natural and understandable it is to want vengeance, surely Christians could agree it's not the right thing to do.

Right?

Well, that's where the interesting dynamics of a Trump-devoted conservative evangelical movement take things to the next level.

In this era, sins are still sins, but they must be weighed against what are perceived to be greater concerns.

In other words, almost anything these days that you would expect Christians to condemn or oppose is not condemned or opposed because, as they see it, there is a greater moral consideration that takes precedence.

Sometimes, they argue, we must choose or excuse evils because the outcome of the culture war is more important both to Christians and to God Himself.

So, is there an argument here?

Some of them look past immoralities when the scale of those immoralities pales in comparison to the concerns of abortion, religious freedom, and the safety of American citizens.

They believe that one should be more concerned with the lives and happiness of their children than whether or not a president is a lying, philandering, unethical charlatan.

If a man, even one with all those attributes, is able to build a wall to protect the citizenry, able to appoint Supreme Court justices who will overturn abortion laws, and able to protect Christians in their right to express their convictions without fear of prosecution or persecution, then aren't the highest moral goods being served, thus making the compromises irrelevant?

We may argue that a higher moral law is being served, and thus be content that it is being upheld by any means, but how can we be so sure that a higher moral law is *being* served?

As Christians, we believe that the future of humanity is in God's hands. But how we conduct ourselves here on Earth has always been a matter of free will. We are not puppets; we are each of us a small part of a larger tapestry. Where we ultimately end up in that tapestry is our choice.

God has shown us what is right. It is up to us to obey or to rebel. But to recast sin as virtue while claiming it is in the service of God? We owe Him a better and more honest effort than that.

We owe it to our nation, our brethren, our neighbors, and the world to get this right.

THE IMMORAL MAJORITY

Chapter 1

THE SHIFT

In the summer of 2016 the idea that the real estate mogul and reality television star Donald Trump might actually become president was still utterly laughable to most experts, pundits, and the Washington power structure. Even at that late date, the belief in his ultimate success was limited mainly to the faithful or the afraid.

Despite having bested sixteen Republican contenders in the primaries, Trump was still viewed by many as a showman. Or, more accurately, a show.

He was at once the carnival barker and the sideshow being barked, and in the circles that his followers would call "elite," he was seen as little more than a public relations savant who had learned the language of the conservative base and spoke it well enough to convince them he was one of them. But none of these doubters thought it would last. Surely people would see through the charade.

Not only did their gut tell them Trump's act couldn't be sustained, there were also plenty of solid reasons to doubt a Trump victory just by looking at the numbers. Every poll was against him. Every news story seemed to reveal a new low to which his character could sink.

And despite enjoying a truly unprecedented level of media saturation, his every word and action afforded wall-to-wall coverage by the morbidly fascinated and ratings-intoxicated press, these

Trump skeptics found some comfort in the fact that the stories were almost universally negative outside of the relatively safe space offered by Fox News.

In the conservative Republican circles where I lived and worked, the feeling was only slightly different. We believed there was little chance he wouldn't be the nominee, despite threats of a brokered convention, but saw little chance of a general election win, which was bittersweet to us, as it forced us to confront the idea that his loss meant Hillary Clinton would be president, an outcome that many of us had spent years working to prevent.

The dread was so potent among that minority of Republicans who refused to get on the "Trump train" that we banded together like a gang and were referred to (and are still referred to by some) as "Never Trump." The name was a call to inaction of sorts. The shared premise among us being that our principles did not permit us to vote for Donald Trump, even if it resulted in a Clinton victory.

But as I said, at this point in the whole sordid affair, most of the strife that existed within the party was about how the next several months would go leading up to the inevitable Republican loss. Arguments about what a Trump presidency would be like seemed too far-fetched to seriously engage in.

Among the reasons already mentioned that Trump could never be president, one simple one was this: Christian evangelicals would never vote for this man, and Trump couldn't win without Christian evangelicals.

It wasn't even about whether or not Trump, as a Republican, might govern in ways that aligned with the community's policy wishes to a greater degree than Democrats could. It was about what made evangelicals a voter bloc in the first place. About what it meant to be an evangelical.

Evangelicals purportedly adhered to the wisdom of Matthew 16:26: "For what does it profit a man to gain the whole world and forfeit his soul?"

Given the number of things that had to be overlooked about Donald Trump's character in order to pull the lever for him that November, this barrier to evangelical support seemed insurmountable.

In retrospect, it's easy to say that I was blind or a purist or even simple-minded, but the clarity of hindsight aside, I simply couldn't see how Christian leaders could rally behind this man.

Political expediency just doesn't comport with the idea of a "higher calling," and all my life, growing up in the home of an evangelical minister and attending church and speaking with other Christians, the "higher calling" was always portrayed as the primary consideration.

I guess I knew at some level that it would require a dramatic shift at the heart of the American church, and I just couldn't fathom such a shift taking place.

And no shift meant no President Trump. Case closed.

We now know, of course, that the case was far from closed.

But back to the summer of that year, there was a particular moment when the scales fell from my eyes and I could see clearly what would happen next.

The bellwether for me was a tweet from evangelical leader Jerry Falwell Jr. on June 21, 2016.

It was a photo that, for me, spoke a thousand truths.

Falwell, the head of premier evangelical Christian school, Liberty University, had endorsed Trump earlier that year. At the time of the endorsement I was far from convinced Trump would even be the nominee, so the gravity of it all just didn't hit me. Not until this tweet and now-famous photo.

"Honored to introduce [Donald Trump] at religious leader summit in NYC today! He did incredible job!" the tweet read. In the photo, Falwell stood shoulder to shoulder with Trump with his thumb extended upward in approval.[1]

Falwell Jr.'s wife, Becki Falwell, was also in the photo, and just over her left shoulder one of the many magazine covers proudly displayed on Donald Trump's wall was clearly and prominently visible. Trump appeared on the cover of all the magazines on display, but one cover was of particular note: the March 1990 edition of *Playboy*, with playmate Brandi Brandt wearing nothing but Trump's tuxedo jacket.

I've used the word *stunned* to describe how I felt in that moment, but let me be clear. My reaction to this display was not a holier-than-thou expression of judgmental distaste. Nor was it a Victorian outburst of prudish shock. If Donald Trump was proud of his *Playboy* cover, that was neither a surprise nor a particular horror.

I felt stunned because it triggered a very vivid memory. That of me as a small child living in Dallas, Texas, in 1984 and being taken to a protest by my evangelical parents. A protest of roughly five thousand anti-pornography demonstrators who objected to the sale of *Playboy* in 7-Eleven convenience stores. A protest led by Falwell Jr.'s more famous father, Jerry Falwell Sr.[2]

Jerry Falwell Sr. had actually been a presence in my life beyond that protest. After my family uprooted from Texas and relocated to Lynchburg, Virginia, my father, Dr. Thomas Howe, attended Falwell's Liberty University to work on earning those two letters that precede his name today.

Because both of my parents worked at Liberty as well, we attended Thomas Road Baptist Church where Falwell Sr., who founded the church in 1956, was the pastor.

Falwell Sr. was widely regarded as one of the most influential evangelical leaders in the country at the time my family sat in those pews, and now, more than a decade after his death, his son had taken on the mantle of continuing his legacy as the president of Liberty University.

And at that moment, with that tweet, it felt almost poetic. Like it was too perfectly scripted to be real.

The son of arguably the most famous evangelical leader of the past thirty years; the son of a man who founded an organization called the Moral Majority, which sought to grow a movement of civic-minded Christians for the express purpose of ensuring issues of morality were part of the political discourse; the son of a man who had fought vigorously and viciously against former President Bill Clinton for issues solely related to character and morality . . . This son now poses with a confessed and apparently unrepentant serial adulterer in front of a proudly displayed magazine emblazoned with Trump's image—a magazine against which Falwell's father had literally marched?

The 1983 book *The New Christian Right*, by Robert C. Liebman and Robert Wuthnow, said of Falwell's Moral Majority's membership, "The most critical assumption behind Moral Majority positions seems to be a strong concern for the negative impact of a trend toward a secular culture on traditional institutions in American society, especially the church, education, and the family." Among other things, the Moral Majority stood in "opposition to secular humanism (i.e., human-centered rather than God-centered morality)."

And of pornography, in particular, Falwell Sr. is quoted as saying, "Pornography hurts anyone who reads it—garbage in, garbage out."

In fact, Falwell Sr. was interviewed by journalists Andrew Duncan and Sasthi Brata in January 1981 discussing how disappointed he was in evangelical Christian president Jimmy Carter for having an interview published in *Playboy*, saying that it "was lending the credence and dignity of the highest office in the land to a salacious, vulgar magazine, that did not even deserve the time of day . . . I feel that he was pitching, he was campaigning to an audience that doesn't read the Baptist Sunday school's quarterlies."

Later, Falwell sued *Penthouse* owner Larry Flynt over the publication of a parody ad that portrayed the preacher in an incestuous relationship with his mother. He lost the cause but ironically formed a friendship with Flynt that would last until Falwell Sr.'s death in 2007. However, it's clear the friendship was not an endorsement of Flynt's lifestyle. According to Flynt, "[Falwell Sr.] wanted to save me and was determined to get me out of 'the business.'" Essentially, Flynt was suggesting that Falwell Sr. was trying *to* influence a pornographer, not to put a pornographer into a position *of* influence.

Yet here was Falwell's son, not seeking to "save" a man who was cut from the same cloth as Flynt but rather to endorse an effort to put such a man into the most influential position on the face of the Earth.

It's not my aim to belabor the tweet or overstate its importance in the world, but as I said, sometimes it is the small things that pull the blinders off. In that moment I really couldn't help but see it as an encapsulation of what appeared to be happening to the evangelical as well as conservative movements: The son departing from the father. The Christian departing from the Christ-like. The conservative becoming the populist.

It felt like I was witnessing a metamorphosis. And this tiny moment, this small tweet, this singular photo captured it all in a heartbeat. The doctrine of evangelicalism as I had known it, one driven by its biblical principles above all other considerations, was either shifting into something different or exposing itself for what it had always been.

As important as the moment was to me, it was only the beginning.

In the Beginning

In the beginning there was . . . a beginning.

If someone has spent their life not believing that there is an all-powerful God who created the universe, it's pointless to argue with them or attempt to persuade them by beginning your argument with the Bible and working from there. You don't start with telling someone who doesn't believe in Jesus how much Jesus loves them.

They don't believe in God. They don't believe in Jesus. So why on God's green Earth would it matter to them that someone they don't believe even exists loves them?

If you're on a blind date, don't start by saying, "I love you, marry me," and expect to work backward from there to "Well, I like long walks on the beach." First of all, you might get maced. And, second, you can't expect a person to be interested in or desirous of the benefit or attention of a person about whom they know—or believe—almost nothing.

Yet, so many Christians do this. They start with "Jesus loves you." Likely because the effect of God in their own lives has been so profound that they have a hard time comprehending how anyone wouldn't want to feel the way they feel. As an emotional appeal, this can work, but if you're arguing with an atheist, he isn't invested in how God feels about him, but rather whether He is even a real thing.

All this is to say that I shouldn't start this book on a presumed shared premise, even though I have an objective and am making a precise case. So it seems worthwhile to back up for a moment and establish not only what I perceived to be mainstream Christian values that had been betrayed by the likes of Jerry Falwell Jr., but also to take an objective look at what the general view of evangelicals had been prior to this marriage of a movement to a man.

Where did I get my idea about what Christian values are in the first place?

I grew up in the home of a pastor and a pastor's wife who, over the course of my life, instilled in me a connection to God and a compassion for people that I have tried to teach my own children. For me, my mother and father were the picture of Christians.

They expected more of themselves than others. They could be strict but not without a measure of understanding. They were devoted to their marriage, their children, and their God. They owned their mistakes and never presented themselves as being better than others.

Sometime in my early teenage years, after I'd been exposed to enough people who held a deep contempt for Christians, I asked my dad, "Why do people hate Christians so much?"

He answered, "Because so many Christians are jerks." His answer stuck with me, but in my naïveté I assumed that his view—which to me spoke to the fact that salvation does not make you better than anyone else—was the common view of Christians.

Even more than Christians, we were evangelicals. My dad, when he had been a preacher, had been a Southern Baptist preacher.

My entire childhood was in the South: Georgia, North Carolina, Texas, Virginia. And we'd attended famously evangelical churches such as W. A. Criswell's First Baptist Church in Dallas,

Texas, and, as mentioned earlier, Pastor Jerry Falwell's Thomas Road Baptist Church in Lynchburg, Virginia.

When we lived in Lynchburg, Virginia, in the early 1990s, my mom worked as a receptionist at Liberty University, a Christian school founded by Falwell, and my dad both taught and attended courses at the same university under the tutelage of biblical apologetics titan, Dr. Norman Geisler.

Outside of my parents, my main Christian influences were my father's colleagues and mentors, like Dr. Geisler, or my uncle, Dr. Richard Howe, a seminary professor himself with a PhD in the philosophy of religion, just like my father.

The men and women who surrounded me when I was growing up had a far greater influence on my understanding of Christianity than the pastors I ignored in church on Sundays. In fact, I recall my dad often being dissatisfied with the pastors of our churches, which, whether his intention or not, caused me to dismiss most of their sermons.

I do wonder how different my understanding of evangelicals might be had I listened more closely.

But despite how little I paid attention to Falwell thundering or Criswell crying, I was still exposed to a great deal of Christian thinking, thanks to my parents and the people who made up our circle, as a result of my father's work.

I was a troubled teen. Despite my parents' raising me to know better, I succumbed to peer pressure and dived deep into the world of drugs and crime. Those problems got only worse after I dropped out of high school and started dealing drugs.

Through it all, my parents struggled to figure out how to control me. Once a teenager of seventeen or eighteen years decides to be rebellious, there is little that a parent can do, no matter how much other parents may swear they "wouldn't put up with it."

A lot of my friends were kicked out of their homes by their parents as a "tough love" measure, something my parents refused to do to me. Looking back and remembering the reactions of the friends whose parents did this, I think my parents were right. Where those friends who were kicked out felt *some* shame, they mostly felt abandoned. I, on the other hand, felt greater shame for my bad decisions precisely *because* my parents didn't abandon me.

In fact, my mom even took in several of my friends whose parents had kicked them out. To the point that friends jokingly referred to my house as the "Howe halfway house." But despite those jokes, any one of my friends whom she took in would defend her as if she were their own mother. I know this because they've told me. She showed them something they needed: unconditional love.

That unconditional love was also clear to me, and the lesson it taught me has impacted my entire life.

Around age eighteen, having been banned from a local hangout (a strip mall in Charlotte, North Carolina, called the Arboretum), I recklessly got dropped off there one night with a breakfast buffet of drugs and paraphernalia on me. It didn't take long for the police, who had already arrested me there once for trespassing, to recognize me and arrest me again.

But unlike my previous run-in with the law, I was no longer a minor. This was going to be part of my permanent record.

When I called my parents from jail, they said they would be willing to get me out—but if I was going to be in their custody, they were going to have certain expectations of me.

As I waited for them I imagined all types of consequences. Never going anywhere. Not being allowed to do anything or talk to anyone. A plethora of chores.

So I was pretty surprised by the consequence my dad explained to me.

I was to invite all my friends, no matter how much of a delin-quent they may be, to my house every Sunday. Not for Bible study, not to cram God down their throats. My dad just wanted to talk. The topics would be up to my friends and me.

That first Sunday we gathered, I pressured my friends into at-tending as a favor to me, and while a handful of them complied, none wanted to be there. There were, perhaps, five of us.

The conversation did touch on religion, but also on morality in general, plus world events, history, relationships. Everything. Dad was patient and attentive. He didn't treat me as more important than anyone else, and he approached every question with serious-ness and a dash of levity.

Within a few months, after the charges against me had been dropped by the county, making me "free" of the consequence, the meetings at our home continued. The gatherings had become a mostly regular thing, and as many as ten or fifteen of my friends were attending them. They looked forward to them and would think of questions they wanted answered days before we'd even gather.

At times, my uncle Richard would also attend, as well as other seminary colleagues of my father. The meetings had become less about asking questions and more like a collaboration. Often, my dad didn't even have to involve himself in a philosophical debate, because those in attendance had started to familiarize themselves with the language and had simply become more curious. And, I'd argue, more wise.

I still know some of these friends, thanks to social media like Facebook, and to this day they will remark on the impact those meetings had on their lives.

Now that I'm in my forties, it's interesting to consider. What is it that these friends gained? Certainly not all of them are Christians, and some are almost certainly atheists. Yet they loved these meetings. Why?

The answer is simple. They felt welcome. Included. Important. Valuable.

Like most people, they had a caricature of the southern Christian evangelical. Judgmental and isolating. Indifferent to others. But my parents turned that concept on its head. They welcomed criminals, individuals they could believe were corrupting their son, into their home. My mom fed them and gave them beds. My dad talked about life and philosophy with them and told them how smart they were.

Growing up in these circumstances could explain why I had been resistant to seeing that which I eventually saw. I held on to an idyllic view of Christians because, I suppose, in a lot of ways my parents are what I'd consider ideal Christians, and ideal Christians were all I'd ever known.

Coming around to the idea that many Christians weren't self-aware sinners who were striving to be Christ-like was something that I obviously took a long time to acknowledge. Recognizing that evangelicals had gone a step further, into the realm of self-delusion and moral superiority, took even longer.

My experience, however, is not universal. While my parents had been a great example of Christian charity, it's safe to say that much of America has a very different view of evangelicals than I did.

Recent events have forced me to stop and consider all of the ways that evangelicals have been portrayed to me over the years. Looking at that photo of Jerry Falwell Jr. and Donald Trump, I thought back on the times I'd defended my fellow churchgoers against the scorn of the secular world, and I had to ask myself a question.

Had all the critics been right?

The Evangelical Caricature

The stereotype that many people have of evangelical Christians has been featured in countless movies and television shows, skewered in news panels, debated at dinner tables, and satirized on comedy shows.

It is an image that is exemplified by the likes of Jimmy Swaggart, a Pentecostal evangelical, who rose to prominence after launching his television ministry, the *Jimmy Swaggart Telecast*, in 1971. By the early 1980s, Swaggart's weekly telecast was reaching millions of homes through hundreds of television stations across the country.

Swaggart was prone to histrionics, including a penchant for weeping at the pulpit, and was known for castigating fellow evangelists for their weak moral fiber following numerous scandals in the televangelist community.

The *New York Times* described Swaggart's moralistic crusades against these "sissified preachers," saying, "The self-styled 'old-fashioned, Holy Ghost–filled, shouting, weeping, soul-winning, Gospel-preaching preacher' would glower through his horn-rimmed glasses and shout for deliverance from 'pret-ty lit-tle boys with their hair done and their nails done, who called themselves preachers.'"

"Sissified preachers" such as fellow televangelist Jim Bakker, who in 1987 had been exposed for spending more than $270,000 attempting to silence a former female employee with whom he had had sex in a Florida hotel.

It would later be revealed that Bakker had that type of money to spend as a result of the $3.7 million he'd embezzled from the $158 million worth of fraudulent vacations he'd sold to his followers.[3]

He would end up serving five years of an eight-year prison sentence for his crimes.

In light of the sex scandal, Swaggart called Bakker a "cancer on the body of Christ"[4] that needed to be "excised."

"The church cannot hide sin," Swaggart said. "When a preacher has been found out and it is a fact—not hearsay—that he has performed an act of adultery, then a hearing is convened and then he has to step down."[5]

But before long it was Swaggart who found himself in the hot seat, confessing to some eight thousand congregants at his World Faith Center in Baton Rouge, Louisiana, in 1988 for what was eventually revealed to be an extramarital sexual encounter with a prostitute.

In fact, he had been discovered only as a result of yet another Pentecostal minister, Marvin Gorman, who in 1986 had been accused by Swaggart of having multiple extramarital affairs, a charge Gorman denied, though he did confess to an "attempted" adulterous act with another pastor's wife years earlier.[6] That confession had been enough to earn him a defrocking from his ministry.

A year later, Swaggart was photographed having a sexual encounter with a prostitute at a seedy hotel in Baton Rouge. After taking the photos, the photographer disabled Swaggart's car and contacted Gorman, who rushed to the hotel and confronted Swaggart in the parking lot, which eventually led to Swaggart's tearful confession and dismissal from his ministry, a ministry that purportedly brought in $140 million per year at the time.[7]

There was also Reverend Robert Tilton, whose ministry Success N Life reached 199,000 households by 1991 and took in $7 million in donations per month, often the result of Tilton's pledge to answer all prayer requests, most of which had been mailed in with cash.

An ABC investigation revealed that the prayer requests had been thrown away by the thousands after the cash had been collected and without Tilton ever seeing them.[8] In fact, tens of thousands of these trashed prayer requests were discovered in the dumpster behind the very bank at which the donations were deposited.

In the following years, Tilton was the defendant in several lawsuits brought by former donors who felt defrauded by his actions, in light of the investigation that exposed the ministry as being little more than a moneymaking scheme.[9]

Another example: televangelist Peter Popoff, whose nationally broadcast show involved the minister using his "God-given ability" of divine revelation to call ailing church audience members to the front, where Popoff would "heal" them.

In 1986, it was revealed that his wife would speak to him through an earpiece to provide information they had collected ahead of the sermon via questionnaire. Audience members who had filled out these questionnaires would be selected by Popoff, who would then use his faux-divine power to correctly identify their prayer requests, and a "miraculous healing" often followed.[10] At the time he was exposed for the cynical scam, Popoff's ministry was raking in millions of dollars per year.

These and many other scandals among some of America's most visible evangelists, combined with the staggering sums of money brought in to these so-called churches, solidified an image—fair or not—of the evangelical.

Figures who were cartoonish, dramatic, deceitful, wealthy, white, smarmy, judgmental, callous, and, above all, hypocritical.

Charlatans.

It is a stigma that has chased evangelicals for decades. Even thirty years later, the memory of these televangelist scandals prompts some evangelical leaders to segregate the Pentecostal tel-evangelists as a doctrinally isolated fringe group.

For example, as recently as January 2016, Russell Moore, the president of the Ethics & Religious Liberty Commission of the Southern Baptist Convention and an outspoken Trump critic, described then-candidate Donald Trump's evangelical support as being from the "Jimmy Swaggart" wing of the movement, explaining that Trump "tends to work most closely with the prosperity wing of Pentecostalism."

His description seemed to serve as much as a diminishment of Swaggart's inclusion in the evangelical movement as it was a minimizing of Christian support for Donald Trump, who, to some evangelical leaders at the time, was cut from the same cloth as Swaggart but without even the false veneer of moral leadership.

But the image of evangelicals was not solely shaped by televangelists of the 1980s and '90s.

Ted Haggard, former megachurch pastor and president of the National Association of Evangelicals, resigned his positions in 2006 after it was revealed that he had paid a male prostitute for sex.[11]

In March 2018, Frank S. Page, the chief executive officer of the Southern Baptist Convention's Executive Committee, resigned his position following what was described as "a morally inappropriate relationship in the recent past."[12]

Memphis megachurch pastor Andy Savage faced accusations that he had sexually assaulted a woman twenty years earlier when he served as a youth pastor in Texas and she was still in high school. According to the accuser, Savage not only had sexual intercourse with her when he was an adult and she was a minor under his care, he also asked her to remain silent about it. She testified that the church they both attended also instructed her to remain silent.

Savage resigned from Highpoint Church after confessing to the encounter and apologizing. For this confession he received a standing ovation from the congregation.[13]

In early 2019, scandal engulfed Southern Baptist churches as an investigation at the *Houston Chronicle* revealed more than seven hundred victims of sexual abuse by church officials over the last twenty years.[14]

The revelations, as summarized by the *Chronicle* in a subsequent report, are nothing short of shocking.

Journalists in the two newsrooms spent more than six months reviewing thousands of pages of court, prison and police records and conducting hundreds of interviews. They built a database of former leaders in Southern Baptist churches who have been convicted of sex crimes.

The investigation reveals that:

At least thirty-five church pastors, employees, and volunteers who exhibited predatory behavior were still able to find jobs at churches during the past two decades. In some cases, church leaders apparently failed to alert law enforcement about complaints or to warn other congregations about allegations of misconduct.

Several past presidents and prominent leaders of the Southern Baptist Convention are among those criticized by victims for concealing or mishandling abuse complaints within their own churches or seminaries.

Some registered sex offenders returned to the pulpit. Others remain there, including a Houston preacher who sexually assaulted a teenager and now is the principal officer of a Houston nonprofit that works with student organizations, federal records show. Its name: Touching the Future Today Inc.

Many of the victims were adolescents who were molested, sent explicit photos or texts, exposed to pornography, photographed nude, or repeatedly raped by youth pastors. Some victims as young as three were molested or raped inside pastors' studies and Sunday school classrooms. A few were adults—women and men who sought pastoral guidance and instead were seduced or sexually assaulted.[15]

Of the ten churches that were identified in the aftermath of the investigation to warrant greater scrutiny from the Southern Baptist Convention, seven were cleared by the convention.

One of the three that remained under investigation by the convention, Bolivar Baptist Church in Sanger, Texas, had been led by a pastor who confessed to impregnating a teenage church member at his previous church while he was the pastor.[16]

And these are merely the scandals that erupted and caused prison sentences, resignations, and disgrace. There are also countless stories that pepper the news every year and continue to show a movement beset by snake oil salesmen.

The image of a man in a fresh suit with a pearly white smile, his hands raised to the sky while promising cures and wealth and happiness even as he rakes in donations that allow him to live a lavish lifestyle is certainly a common one for evangelical leaders and pastors.

Evangelicals seeking to combat that image, which they see as an unfair generalization of a diverse faith coalition, often reference the biblical lessons of redemption, repentance, and the fallen nature of man.

The idea is that no one can live up to God's standard for us, even His most devout followers. Basically, the message is this: the fact that Christians sometimes behave as hypocrites is actually proof that they're not! After all, if evangelical pastors are found banging their fists on a pulpit while declaring the despicable sinful nature of human beings, how can it be hypocritical of them actually to *be* sinful human beings? They said they were right there in the sermon!

And of course that's when repentance comes in, because these evangelical pastors would also acknowledge that, regardless of our sinful nature, we must confess and seek forgiveness.

It's all very neatly put together with a faithful bow tied on top. In fact, as a Christian myself, I acknowledge that they aren't even wrong. It's true that among all walks of life, all faiths, all ideologies and philosophies, every nation, every creed, every gender, and every age group, we are all subject to temptation and sin, and at one time or another, we will give in to temptation. And when we do, we must learn from our mistakes, seek reconciliation and forgiveness, and undertake to better ourselves while accepting whatever consequences may come from our actions.

And frankly, for years, this defense was good enough for me. I saw who Swaggart was as much as I saw who Bakker was. I knew where the caricature of evangelicals came from as much as I knew that being true to myself and my convictions was more important than whether or not outsiders chose to group me with bad actors in the movement.

Different people have different moments in their lives where they find their eyes opened to something that had been in front of them all along. How it comes about and what triggers it could be an enormous life event or a tiny moment that symbolizes something larger.

"So many Christians are jerks," my dad said all those years ago, and as I said, it stuck with me. But that wasn't the moment when my eyes opened. That moment occurred years later, during the 2016 election.

Before the Shift

June 2016 was an important month in Donald Trump's campaign momentum for reasons far beyond the photo of him with Jerry Falwell Jr.

This was in part because, prior to that moment, the primaries were still very much a contest. While in hindsight we may see Trump's rise as an unstoppable force, the truth is that in the months ahead of that summer, things were not so certain. Attitudes, even among Christians, were divided.

During that time, aside from the almost universal disbelief that Trump could ever win, there existed still a sizable portion of the evangelical Republican base that believed he was a bad representative for their values. That portion had a public voice.

Darren Patrick Guerra wrote in March 2016 that in the Super Tuesday races on March 1, "Trump failed to win a majority of evangelicals in any southern state and lost more than half of evangelicals, on average, overall. A look at the second Super Tuesday from March 15 reveals similar results with a couple of surprises. The bottom line is that a majority of evangelicals are still backing candidates other than Trump. In Missouri, the most religiously active voters are supporting non-Trump alternatives with numbers as high as 70 percent."[17]

Among Christians more likely to regularly attend church (a feature primarily attributed to evangelicals as a subset of American Christians), the lack of support for Trump seemed very real and clear in *early* 2016. While Trump enjoyed a 32 percent share of Republican Christians in general at the time, that number dropped to 24 percent when limited to Christians who had attended church in the week prior to being asked.

Even many evangelical leaders were publicly lamenting the levels of approval he was receiving from ostensibly "value-driven" voters.

Pastor Max Lucado of San Antonio, Texas, said in an editorial for the *Washington Post* in February 2016 that he was "chagrined" by Trump's antics.

> *He ridiculed a war hero. He made a mockery of a reporter's menstrual cycle. He made fun of a disabled reporter. He referred to a former first lady, Barbara Bush, as "mommy" and belittled Jeb Bush for bringing her on the campaign trail. He routinely calls people "stupid" and "dummy." One writer catalogued 64 occasions that he called someone "loser." These were not off-line, backstage, overheard, not-to-be-repeated comments. They were publicly and intentionally tweeted, recorded and presented.*[18]

Lucado went on to question how Christians could support a man doing these things as a candidate for president, much less as someone who repeatedly attempted to capture evangelical audiences by portraying himself as similarly committed to Christian values.

He continued, "If a public personality calls on Christ one day and calls someone a 'bimbo' the next, is something not awry? And to do so, not once, but repeatedly, unrepentantly and unapologetically? We stand against bullying in schools. Shouldn't we do the same in presidential politics?"

Rolling Stone reported on several evangelical leaders pushing against a Trump nomination, including North Carolina radio host and evangelical Dr. Michael Brown, who wrote an open letter to Jerry Falwell Jr., blasting his endorsement of Donald Trump. Brown wrote, "As an evangelical follower of Jesus, the contrast is between putting nationalism first or the kingdom of God first. From my vantage point, you and other evangelicals seem to have put nationalism first, and that is what deeply concerns me."[19]

John Stemberger, president and general counsel for Florida Family Action, lamented to CNN, "The really puzzling thing is that Donald Trump defies every stereotype of a candidate you would typically expect Christians to vote for." He wondered, "Should evangelical Christians choose to elect a man I believe would be the most immoral and ungodly person ever to be president of the United States?"[20]

A May 2016 report from Katie Zezima at the *Washington Post* highlighted how the apparent split among evangelicals was even finding husband pitted against wife.[21]

The fissure is playing out in some households, including Rich and Heather Dreesman's in Plattsmouth, Neb.

Rich Dreesman doesn't like Trump, calling him "not a godly man" and "kind of a lunatic." But he will probably vote for him in November because he believes Democrats and Hillary Clinton are "evil" and "ignorant." His antipathy toward Democrats is strong: He said he wanted to write into his will that none of his five children would receive their portion of his estate if they registered as Democrats; he fired his lawyer for saying no.

Heather Dreesman said she is diametrically opposed to Trump on a long list of issues, including transgender bathrooms and his tax and immigration policies, and believes he will not protect religious freedom. She finds Trump crass, vulgar and a misogynist.

"As a conscientious believer, I just can't vote for someone who supports some of his philosophies," she said. "I think he doesn't know what it means to be a Christian."

Russell Moore, president of the Ethics & Religious Liberty Commission of the Southern Baptist Convention, was so disgusted with evangelical support of Trump that he claimed to have stopped referring to himself as an evangelical for the remainder of the campaign season, saying, "When this fevered moment is over, we will need to make 'evangelical' great again."[22]

Moore identified Trump as "an unrepentant serial adulterer who has abandoned two wives for other women," adding, "I don't think this is someone who represents the values that evangelicals in this country aspire to."[23]

Moore was even called to task over his remarks by Trump himself, who tweeted in response, "Russell Moore is truly a terrible representative of Evangelicals and all of the good they stand for. A nasty guy with no heart!"[24]

Most of the objections to Trump at this time—which should be noted were all prior to the June shift I've described—had to do with the feeling that he could not be a leader who would bring Christian values to the Oval Office because he simply didn't understand them and quite often seemed to be selling the idea that he was faithful the way he might sell Trump Steaks.

When speaking at Liberty University just weeks prior to the Iowa Caucus, Trump's salesmanship as a Christian was on full display. At that speech to an assemblage of Liberty University students, who were required to attend, he read from his notes, including a Bible verse from 2 Corinthians 3:17, "Now the Lord is that Spirit: and where the Spirit of the Lord is, there is Liberty."

But in what many took as a lack of familiarity with basic Christian parlance, Trump read the book and chapter as "Two Corinthians" instead of the more common "Second Corinthians."

Honestly, it was probably just a simple mistake that anyone could have made when reading from notes that had been written by someone else (the notes had been provided by evangelical Tony Perkins, president of the Family Research Council). However, because Trump was already perceived to be falsely selling himself as a fellow Christian just to win the election, the moment seemed like an apt example of his shallow faith.

To be clear, questioning the truth of another person's faith is not an action Christians should usually or casually undertake. However, it is also true that one must be vigilant against deception. It does seem worth noting that, regardless of the health of his private relationship with God, Trump's devotion to understanding the tenets of faith appeared Cliff Notes–level at best.

Trump's repeated claims that he'd been a church-going Protestant his entire life painted a picture of someone who may very well have clocked in and out of "God time" but also someone who had little genuine interest or curiosity to learn much more than could be picked up by reading a Wikipedia page. Even non-Christians who have been around Christians know some of the basics. It appeared this wasn't true of Trump.

When it comes to Christianity, Trump knows the notes but not the tune. When Trump was asked if he'd ever entreated God's forgiveness, the surrounding crowd laughed at the question. They knew it directly challenged Trump's public image as a man who never admitted he was wrong. Trump said, "I'm not sure I have." He went on, "When we go in church and when I drink my little wine . . . and have my little cracker, I guess that's a form of asking for forgiveness, and I do that as often as possible because I feel cleansed." His words exposed his ignorance of both the solemnity of the ritual and its actual theological meaning as an expression of Christ's ultimate sacrifice.

What remains depressingly fascinating about Trump is how concerns over his lack of familiarity with or his authenticity on

Christian issues are often best illustrated in the way he rejects those very concerns.

For instance, after Pope Francis was quoted as saying, "A person who thinks only about building walls, wherever they may be, and not building bridges, is not Christian," most outlets reasonably concluded that he was making reference to Trump's campaign centerpiece: a wall at America's southern border with Mexico. Trump evidently concurred that it was directed at him, and he hit back, saying to a crowd of supporters in South Carolina, "No leader, especially a religious leader, has the right to question another man's religion or faith."

Yet, at CNN, they compiled a full history of Trump's calling Barack Obama's faith into question, including saying of Obama's birth certificate, "He may have one, but there's something on that, maybe religion, maybe it says he is a Muslim," as well as tweeting in 2012, "Does Madonna know something we all don't about Barack? At a concert she said 'we have a black Muslim in the White House.'"

That same year he also tweeted an article from the conspiratorial website World Net Daily, suggesting a ring President Obama was wearing indicated some sort of Arab origin, saying, "Why does Barack Obama's ring have an [A]rabic inscription," and, after linking the article, adding, "Who is this guy?"

It was precisely this type of easily identifiable and seemingly shameless false statements and hypocrisy that continued to make support for him a point of confusion among dissenting evangelicals. But also, sadly, it was a feature that attracted others.

As I said earlier, for Christians like myself and some of those quoted here, it was hard to imagine how evangelicals could get truly excited for Trump . . . unless we were simply wrong about what broadly appealed to evangelicals.

Guess what?

Despite Trump's shaky start in the Iowa Caucus (Ted Cruz won the caucus as well as the evangelical vote), the subsequent primary contests demonstrated that the real estate magnate knew far more about how to appeal to masses of evangelicals than the more traditional evangelical candidates who "spoke the language."

At a rally in Iowa in January 2016, Trump told his audience, "Christianity will have power," if he was elected. "Because if I'm there, you're going to have plenty of power. You don't need anybody else. You're going to have somebody representing you very, very well. Remember that."

This appeal to powerlessness proved to be devastatingly consequential. It was powerlessness that people felt in Iowa and South Carolina. It was a sense of disenfranchisement that was motivating folks in Texas and Mississippi.

For a great many evangelicals, the word *power* was exactly what they wanted to hear. The idea of having it, and wielding it, was more than intoxicating. It was a lifeline. And in the end, it's what many absolutely believe Trump as president has given them.

The Shift

Trump's acceptance among evangelicals continued to increase as summer approached and his nomination moved toward inevitability. Dissenting evangelicals like myself were faced not only with the looming reality but with the uncomfortable truth that evangelicals had helped put him there.

As *Christianity Today* reported at the time, "Even before the other candidates left the race, Trump was the most frequent choice of all evangelicals, including frequent attendees. Although the levels of support changed according to frequency of attendance, he was still the most frequently chosen in every breakdown in most states."[25]

The summer had arrived and the shift had taken place, but the questions, the important questions, are: What happened that precipitated or enabled that change? What took the evangelical movement from questioning and doubting to Jerry Falwell tweets and religious fervor for secular pragmatism?

Again, it's utterly essential to keep in mind that until this shift, the doubters and the uncertainty were public and acceptable. And among Christian groups as a whole, as opposed to specifically evangelicals, the acceptance of Trump as "one of us" or even a worthy secular leader was still very much a debatable prospect.

In other words, you could still offer dissent or rejection of Trump without being branded an apostate.

For instance, after Ted Cruz dropped out and Trump's nomination became certain, the Catholic organization Catholic Vote said the next day in a release, "We will not endorse Donald Trump for President at this time. As much as we oppose Clinton, Trump remains problematic in too many ways to receive our endorsement. With a suspect record, no clear guiding principles, and a history of unpredictability, all we can do for now is take him at his word and hope for the best."

This was the kind of statement I would have expected, and I think a great many observers did expect, from most Christian groups. Broadly, it was present and a problem for the campaign.

Trump's need to gather support from protestant evangelicals, Catholics, and values voters of all stripes had become increasingly important as the transition to the general election began. This was when he needed the big shift that, we now know, he was able to create. Prior to the summer of 2016, it was by no means assured, or at least was not apparent.

In an effort to secure that critical support, candidate Trump held a private meeting with a thousand value-centric conservative leaders in June.

Jerry Falwell Jr., whose endorsement in January 2016 had helped identify him as one of Trump's more enthusiastic evangelical surrogates, had this to say: "If we all wait for the perfect candidate who has the demeanor of our pastors and agrees with us on every issue, including our personal theological beliefs, then we may all sit at home on election day for the rest of our lives. How many of you can honestly say that you even agree with your spouse on everything?" That was the frame, and that meeting was when the evangelical attitude toward Trump transitioned from "peculiar to a subset of Christian voters" to "the way things are."

The importance of the meeting was lost on no one. Penny Nance clarified the purpose and significance, saying of Trump, "He needs to win our hearts."

Evangelical figure Michael Farris, chairman of the Home School Legal Defense Association, who was not invited to the meeting, offered a very pessimistic view. He had been specifically told by one of the meeting's organizers that he was not invited because he'd been too "anti-Trump" up to that point. On the day of the event, Farris had an op-ed in the *Christian Post* in which he referenced his participation long ago at the very first meeting of Falwell Sr.'s Moral Majority back in 1980. He made a remarkable point:

The premise of the meeting in 1980 was that only candidates that reflected a biblical worldview and good character would gain our support. Today, a candidate whose worldview is greed and whose god is his appetites (Philippians 3) is being tacitly endorsed by this throng. They are saying we are Republicans no matter what the candidate believes and no matter how vile and unrepentant his character.

Seeing that stinging rebuke in print now, two years later, it's almost inconceivable that it came from a prominent evangelical, so thorough was the assimilation to come.

"In 1980 I believed that Christians could dramatically influence politics," Farris continued. "Today, we see politics fully influencing a thousand Christian leaders. This is a day of mourning."

In a generation, the movement had changed, he was saying, from trying to be a force for change *in* politics, to being forcefully changed *by* politics.

At the meeting, the endorsements of Trump by evangelical leaders had a clear influence on those churchgoers who might have remained on the fence.

And yet, some Christians at the meeting seemed persuaded by what Trump had to say. "I liked the fact that leaders that I have great respect for—Mike Huckabee, Dr. Ben Carson, Ronnie Floyd, David Jeremiah, Dr. Dobson, many of those who posed questions today—they seemed very pleased with what they're hearing, not only on the floor today, but behind the scenes," [evangelical Ann] White said.

"Hearing these men, who I really respect, testify to the fact that they really believe that Donald Trump is a Christian—maybe not of the kind you and I are, or maybe, they didn't say this, not as far along in his development as a Christian . . . I think that's a pretty bold statement," said the Christian radio host E. W. Jackson in his call. Several Christian broadcasters joining the call with Jackson emphasized the one question on their minds at the meeting: Is Trump really a Christian?[26]

These Christian leaders hoped to push Christians in favor of what amounted to a form of pragmatic faith. The idea being that no matter how much a vote for Trump seemed to defy conventional evangelical thinking, God knows what he's doing.

Ben Carson described God as a chess player moving Trump across the board, saying, "Sometimes he uses a pawn; sometimes he does things in a way that is not very apparent to us. And that's where faith comes in."

Franklin Graham offered more assistance in this attempt to spin pragmatism as faith. NPR reported that Graham told the crowd how "in lots of stories in the Bible, people messed up. After all, Graham told the crowd of devout Christians, the prophet Moses led his people out of slavery in Egypt but disobeyed God; King David committed adultery and murder; and the apostle Peter, who, as one of Jesus' closest followers, really should have had his back, denied three times that he'd ever known Jesus."

Graham went on to say, "There is none of us that are perfect," adding, "There's no perfect person—there's only one, and that's the Lord Jesus Christ, but he's not running for president of the United States."

For Trump's part, in addition to offering the idea that he subscribes to Christian values in his own life, he offered a fair amount of fear as a persuader.

"The next president is going to be very vital . . . in freeing up your religion, freeing up your thoughts," he warned. "You really don't have religious freedom."

Playing to every fear, he also referenced IRS rules governing institutions of faith, saying religious leaders were "petrified" to discuss them for fear of being targeted. He assured the audience, "We are gonna get rid of that."

In keeping with a pattern of assuming the mantle of savior, Trump reminded the audience of Christians, "You can pray for your leaders, and I agree with that. Pray for everyone. But what you really have to do is you have to pray to get everybody out to vote for one specific person. We can't be, again, politically correct and say we pray for all of our leaders, because all of your leaders are selling Christianity down the tubes, selling evangelicals down the tubes."

It's useful to pause and reflect on this.

Christians should care about who will be the president. They are citizens. But what Trump said at that meeting was far more than an appeal to a partisan aim. He was offering himself as God's preference, not just man's, and saying that the Christian thing to do was to pray for him and, ultimately, not to pray for his opponents. His enemies, one might say. And on top of that, he was suggesting that the idea one *should* pray even for one's opponents is lip service of the politically correct kind.

That is incredible to me still.

All of this added up to something far different from Nance's characterization that Trump needed to "win [their] hearts." Instead, it was more like he was there to tell them to get in line. Trump was telling the assembled Christians unequivocally that he was the only Christian choice. That only through voting for him were they serving God. He had his faithful evangelists stand up and profess the same on his behalf. His witnesses.

In response, Trump received a standing ovation at that meeting. Two, in fact. One before he spoke, and another after.

It was a win for Trump. Not because he went in to win hearts, but because he went in ready with a new faith, and with pastors to preach it. And his plan worked.

That was the moment the shift took place.

If your question is, When did the dam break among evangelicals?, the answer is, June 20, 2016.

That's not to say the meeting answers the full question of why, but it is certainly the moment the change was realized.

After that, it was easy for Trump on the religious right. Leading up to November, his popularity among evangelicals continued to rise, thanks in no small part to the efforts of the Christian leaders who worked to build biblical rationalizations that would permit rank and file Christians who had previously wrestled with doubts regarding Trump's character and honesty to vote for him with a clean conscience.

The Trump Evangelicals

And so the Trump evangelical was born.

Having built bridges to mainstream evangelical leaders, having promised a seat of power to grassroots Christians, having even attended and been prayed over by charismatics and fringe groups, Trump had lined up everything he needed to secure the evangelical vote, which now approved of him by a wide majority.

This deliberate strategy was absolutely effective. All told, evangelicals reportedly composed 26 percent of the electorate in the election. And of white self-identified evangelicals, a record-breaking 80 percent voted for Donald Trump.

Ralph Reed, founder and chairman of the Faith and Freedom Coalition, spoke to the *Washington Post* the day after the election about the power that evangelicals had displayed at the polls. Reed described how the aggressive pursuit of evangelicals by Trump and some GOP candidates caused them to be "richly rewarded with a huge faith-based vote that was an indispensable key to their astonishing victories."

Speaking to doubters of evangelical influence, Reed added, "Even as some wrote the premature political obituary of the conservative religious vote, it once again showed its potency and effectiveness. These voters are ignored or spurned by candidates of either party to their great detriment, and likely their defeat."

Trump the evangelical had succeeded, and the Trump evangelicals were here to stay and, effectively, had become the voice of the movement as a whole.

Naturally, then, on election night, many of the heavy-hitting evangelical leaders rejoiced.

First Baptist Church Dallas pastor Dr. Robert Jeffress, a member of Trump's Evangelical Advisory Board, celebrated his victory by telling *Time* that "no Republican candidate has made greater effort to reach out to evangelicals than Trump."

In a sense this was correct. Certainly no one had courted them so deviously, so mechanically, or so relentlessly. An effort to "reach out" isn't how I would characterize it, but the notion that the Trump campaign worked for their vote is accurate.

Jerry Falwell Jr. went further, saying, "When you look at the issues, he ended up being the dream candidate for conservatives and evangelicals. The evangelical community has not been divided on Trump, just the leadership—the people were smarter than their leaders."

In reality, rank-and-file evangelicals, contrary to Falwell's assertion, did vary. It was evangelical leadership that coalesced behind Trump that June, and the flocks then followed.

Peter Leithart, president of the Theopolis Institute, conceded to *Christianity Today* that "Trump is sexually immoral," but he offered hope by adding, "He'll probably guard religious freedom. For that, I woke up on November 9 surprised to find myself relieved, grateful."

President of the Family Research Council Tony Perkins, who endorsed Trump's candidacy, said on Twitter that Trump's victory "provides a much needed opportunity to get America back on track again."

Former Arkansas governor and pastor Mike Huckabee, who in previous presidential primaries had positioned himself as a "faith-based candidate," also took to Twitter and said of Trump's win, "Those who prayed for Trump to win must pray for him to lead with dignity, justice, and strength. The job ahead—YUUGE. But God is able."

Author and radio host Dr. Michael Brown, who earlier criticized Jerry Falwell's endorsement of Trump, wrote that he believed Trump's victory was accomplished by the "sovereign intervention of God."

"Just think of the obstacles Trump overcame," Brown said, as he goes on to list several factors that should have ensured Trump's downfall, including the *Access Hollywood* tape (which he acknowledges as vulgar and "tremendously offensive"), as well as noting Trump's "inappropriate remarks" and the accusations of sexual assault that have been leveled against him. To Brown, the victory in spite of so many attributes that are ostensibly offensive to Christian values is somehow proof that God intervened on Trump's behalf.

Even from those who had at one time offered cautious concern about a Trump presidency, a sense of excitement had taken over, rooted in the idea that Trump's victory must be a reward or a blessing of some sort.

Throngs of Trump-supporting Christians told me throughout the months leading up to the election and on into his presidency that "God has got this."

The overall theme from so many who were unable to deny the unorthodox popularity of Trump among Christians was that we live in a fallen world and must do the best we can within that construct. In the end, many Christians shrugged their shoulders and cast it as a "choice between the lesser of two evils." And, in such an instance, to try to remember that God will work all things for good.

The primary issue I have with this is that Christian leaders simultaneously made clear over and over that a Hillary Clinton presidency would somehow not follow this same reasoning.

If Hillary Clinton were to win the election, we'd be cast into a thousand years of darkness, according to these same folks who now swore that there was no reason for anxiety with President Trump.

"Faith" among the faithful seemed to exist only insofar as it comported with their expectations. Listening to many who expressed this fear of a Clinton presidency, you would be led to believe that God was completely powerless unless we voted the "right" way.

That is the nature of partisanship. Partisanship is the lifeblood of politics. And politics has swallowed at least the evangelical movement whole.

The Christian right, which in the late 1970s had packed its bags and headed to Washington, saying it wanted to bring Christ to government, had found a "seat at the table." And apparently it was a very comfortable seat.

It may be a noble mission to seek to influence the elected through the mobilization of Christian voters who demand the character and values that reflect Christ's teachings. But that is very clearly no longer the mission.

The Moral Majority did not bring Christ to Washington over these past thirty years. Yet for a mission that has, from my vantage point, so clearly failed, they seem quite sure it succeeded.

But whether or not they succeeded may depend on what they were seeking in the first place.

Returning to that photo Jerry Falwell Jr. tweeted in June 2016, given Falwell Sr.'s history of moralizing and loudly condemning any movement or person who failed to meet a baseline of decency, it seemed at first as though Falwell Jr. had betrayed his father's mission. This is not to say that even at that moment I necessarily found my former pastor to be the upstanding example all evangelicals should emulate. He was often a source of frustration for me.

For instance, just after the attacks of September 11, 2001, Falwell Sr. made headlines for an infamous interview with fellow evangelical Pat Robertson on his talk show, *The 700 Club.*

Of the attacks that claimed nearly three thousand innocent American lives, Falwell said, "What we saw on Tuesday, as terrible as it is, could be minuscule if, in fact, God continues to lift the curtain and allow the enemies of America to give us probably what we deserve."[27]

After Robertson concurred with the sentiment, Falwell went on to add, "I really believe that the pagans, and the abortionists, and the feminists, and the gays and the lesbians who are actively trying to make that an alternative lifestyle, the ACLU, People for the American Way, all of them who have tried to secularize America, I point the finger in their face and say, 'You helped this happen.'"

Falwell Sr.'s life may have been free of the sex and financial crime scandals of the televangelists that marked the 1980s and '90s, but the type of bombast steeped in condemnation and lacking in empathy that he displayed in that interview was certainly his contribution to the caricature of the evangelical.

I had disliked much of Falwell Sr.'s approach, but I never doubted whether he believed what he was selling, even if I often thought he sold it poorly. This gave me loads of room for letting things slide, and yet somehow it never occurred to me prior to the shift that perhaps Falwell Sr.'s goals were as worldly as those televangelists.

Recognizing moralistic crusaders as hypocrites was easy when they reached for the low-hanging worldly fruit of sex and wealth, but power, sought for its own sake, is just as much of a vice.

Jerry Falwell Sr. had expressed higher aspirations for his Moral Majority than just the voter mobilization. He sought to leverage those mobilized voters to earn the very seat at the table of power in which Trump evangelicals are currently overly comfortable.

All those years ago, when Falwell made his comments about September 11, I remember feeling revulsion at how callous they were. I felt he should apologize for the remarks, and in fact he did.

But once the Trump evangelical movement had taken hold, I couldn't help but think back to those audiences accepting tearful apologies from the likes of Swaggart and Bakker and how I had viewed them as suckers.

I thought about how over my life, when people would attempt to conflate evangelicalism with these charlatans, I would rationalize that these televangelists were simply "bad actors." That they took advantage of a movement composed of otherwise well-meaning Christians that were, at their core, like what my parents had shown themselves to be.

Eventually, I wondered, Had I fooled myself? Had Falwell simply been my Swaggart? Had I been the very sucker I had been looking down on? Was I too accepting of apologies from charlatans whose goals were worldly and whose stated motives were fraudulent?

The evangelical movement had billed itself as seeking that seat on behalf of the interests of God, and, for most of my life, the judgmental attitudes coming out of evangelical leaders like Falwell Sr. had seemed to support the notion that the power they sought was in the service of a higher moral expectation of this nation and its leaders.

I could disagree with them all I wanted about how they went

about this effort, but it goes beyond disagreement over tactics when a moral crusade sheds the very morality for which they were crusading.

If the movement had now embraced a leader who was a living repudiation of the moral expectations they'd founded the movement to pursue, what would that say about the movement? That it is, was, and had always been about power. Perhaps Falwell Sr.'s son wasn't so far off from his father after all. Perhaps it was not Falwell Jr. betraying the movement his father built.

All of this was eye-opening, of course. But if you had told me that in the two years that lay ahead, the Trump evangelicals would overtly alter their entire ethical framework to accommodate one Republican president, I would still, even after all this, have claimed that was far-fetched.

Yet, that's exactly what's happened.

Chapter 2

THE NEW
GOOD NEWS

My friend once told me about her experience talking to her uncle, a conservative, during the 2016 election cycle. Her uncle had determined that he would not be voting for Trump for president, as he found his character despicable and believed he was unfit for office.

At the time of their conversation, there had been a flurry of news reports about the cascade of evangelical leaders who had stepped up to offer their unwavering support of Trump's candidacy. As such, it was hard for my friend not to wonder how her uncle—whom she described as more of a casual believer as opposed to a devoutly religious evangelical—seemed to hold higher expectations of moral leadership than the Christian leaders whose entire movement had been predicated on the pursuit of those same expectations.

"How do they justify this?" she asked.

"You do not question the vessel that brings your deliverance," he replied.

In the years since the election, this characterization of Trump's evangelical support has proven to be foundational to the theology that has surrounded his presidency.

Are they right? One could argue that the premise is not without biblical support. God chose Noah, a drunkard, to build the Ark. He chose David, a murderer and adulterer, to be king. He chose Saul—who Jesus Himself acknowledged was a persecutor of Christians—to become the apostle Paul.

And it's not the case that in each of these as well as in other instances God's chosen vessels went on to be the stellar examples Christians should want to emulate.

The story of Jonah is often told as a way of expressing that you cannot run or hide from God's purpose for you, but it is also a story of mercy. God's mercy for Jonah. And it was mercy that immediately afterward Jonah could not find for the city of Nineveh, which he thought God should have destroyed. Jonah literally didn't learn the lessons that his own story teaches (Jonah 4:9–11).

Stories of sinners and murderers, thieves and adulterers, litter the Bible, with only Jesus portrayed as living a perfect life. As Proverbs 21:1 says, "The king's heart is a stream of water in the hand of the Lord; he turns it wherever he will." We don't know God's plans. We don't know which individuals God will use, much less *how* they will be used.

So if God's intent is to use this man Donald Trump to expand His kingdom, there are certainly biblical precedents that He may do precisely that. How will He do it and what is His purpose? Those questions are quite separate from determining whether doing it is within His power.

But it is through the theologically sound premise that God can use anyone to accomplish His will that "Trump as God's vessel" has gained steam these last few years. Instead of noting that God often uses people He does not (and we should not) endorse, Trump evangelicals took it upon themselves to canonize Trump, claiming a divine approval for him that he'd never claimed for himself.

Whereas by the end of the election in 2016 the faithful expression was "God can use anyone," the first two years of his presidency have seen Trump become something of a prophesied figure to Trump evangelicals.

In observing and interacting with Christian Trump supporters, this "vessel theology" is something I've encountered a lot. Primarily, the theology exists as a way of explaining how a Christian who enthusiastically supports this president can look away from his many faults, focusing instead on his purpose—a purpose that is often described by Trump evangelicals in near-holy terms. Now, don't think for a moment that I'm unaware of how often leaders, especially presidents, can engender this type of response from folks who are in desperate need of leadership and the restoration of their belief in the promise of America.

I repeatedly encountered more than my share of Obama-devoted activists in the early 2010s on the front lines of the Tea Party. And it's true that many of the progressive adversaries who showed up to counterprotest our efforts seemed to find no possibility of error in President Obama's governance. That level of devotion on its own is not unique.

You certainly will also see such worship lavished on rock stars, famous actors and actresses, and scores of other celebrities who command the seemingly unending adoration of their fans. Politics is no different, except that the impact of scores of such supporters can fundamentally alter the course of a nation.

But there is something remarkably different in the Trump era, and something disturbing and unsettling about Trump evangelicals in particular. Unlike so many other devout followers of a cult figure or celebrity, Trump evangelicals have taken this earthly object of their adoration and quantum-locked him to God's will.

Evangelical Christians are regularly one of the largest voter blocs in American politics. In the last four elections they composed

26 percent of all voters, with a whopping 74 to 80 percent choosing Republicans.[1,2] Their devotion to the Republican Party had always been billed as obedience to a set of Christian values. And members of the GOP were happy to market themselves as the political party most dedicated to protecting Christian interests.

But in 2016, that dynamic became complicated. The apparent character and documented conduct of Donald Trump no longer permitted a blanket statement of "voting our values" when it came to GOP support.

Whereas previously evangelicals may have cited the good moral character displayed by Ronald Reagan or George W. Bush, Trump presented them with a unique challenge. They needed to make the case that they voted for their values when their chosen candidate openly and unapologetically defied those very values.

At first, they found the needed rationalizations to pull that lever, but they kept their words focused on what God can do. Not what Trump can do, per se. They certainly seemed to believe that God could use him to do great things, but there wasn't yet a sense that Donald Trump himself was inerrant. But even once there was, it was a new kind of inerrancy. It's not that these supporters can't acknowledge that something Trump did or said is sinful. It's that when he errs, they seem to feel obligated to shield him from consequence in service of the greater good they imagine he serves. He is inerrant in the sense that he is free of the burden of accountability.

After all, he is the vessel.

Holy Mission

Certainly, as I've demonstrated previously, many Trump evangelicals proclaimed that Trump's victory was in itself a work of God.

On that fateful Tuesday in November 2016, televangelist Jim Bakker, covering the election live, declared that Trump's victory was nothing short of "the greatest miracle [he had] ever seen."[3]

Franklin Graham said of Trump's successful bid for the presidency, "I could sense going across the country that God was going to do something this year. And I believe that at this election, God showed up."[4]

And while it's certainly not unheard of for Christians to greet what they see as good news as evidence of God's plans, there does seem to be something prophetic in the tone of many Christian supporters of this president, especially among those who describe themselves as evangelicals.

A dogmatic law of sorts seems to have taken hold. One that treats Trump support as synonymous with God's will. The standards of what constitutes heresy in this doctrine would make a medieval ecumenical council blush. Trump's governance, and his personal life, have been consecrated.

Consecration was required because, again, the claim is not that Trump himself is inerrant. He is not *inherently* holy. He was *made* holy for this purpose.

Because he's God's vessel. And, as was said, you don't question the vessel.

In 2015, Trump supporter Vicki Sciolaro outlined it on CNN, saying, "God can use anybody. He used the harlots. It's all about what God can do. God can do this. God can use this man."[5] At the time, it was a somewhat common evangelical take on trusting Trump.

But after besting Hillary Clinton, the view shifted for many from speculative to definitive. It was no longer a question of whether God *could* use this man. His election was proof that God *would* use this man. And his supporters were, importantly, quite certain *how* he would use this man. Which is central to the fallacy of their devotion.

In August 2018, Pastor John A. Kilpatrick in Daphne, Alabama, urged his congregants to pray for the president, claiming that he'd received a message from God and that "the Lord said, 'pray for [Trump] now,' because there's about to be a shift and the deep state is about to manifest and it's going to be a showdown like you can't believe."

Kilpatrick went on to speak in tongues and urge his congregation to their feet to pray with him for the president, passionately pleading to heaven, "God make him bold, make him strong! Preserve him, Holy Spirit! Keep him, Holy Spirit! Preserve him, Holy Spirit! Don't let him lose his voice! Make him stronger than ever, Holy Spirit! Let no weapon be formed against him that will knock him out of power. Strengthen him, Lord! It's time to pray, church. I believe our nation is in the balances." [6]

There's nothing wrong with praying for our president. I encourage any Christian to pray for the president, regardless of whether they support him; I make sure to pray for President Trump, despite my feelings about his presidency.

But referencing "the deep state"? Is there any question that when a pastor is praying that the deep state won't prevent Trump from draining the swamp, the marriage of partisan politics and religion is complete, at least in this particular church?

Were that this was the only church.

As a conservative myself, and one who described himself as an evangelical Christian his entire life, it is difficult not to have reached the conclusion that Trump is less of a vessel to these people and more of an idol. It has gotten to the point that one wonders if Jesus Himself would be branded as a purveyor of "fake news" if He were to come down from heaven and condemn an action or utterance by President Trump.

If you think that scenario is outlandish, consider Mark Lee, a member of a focus group on CNN in 2017; he said: "If Jesus Christ gets down off the cross and told me Trump is with Russia, I would tell him, hold on a second, I need to check with the president if it is true. That is how confident I feel in the president."[7]

But who knows? Maybe Mark Lee isn't a Christian. Or maybe he was simply being hyperbolic. It's possible. Perhaps even probable. Even his fellow focus group members seemed shocked at his declaration. More important, Lee has no substantial influence with American Christians. No evangelical gravitas.

But the same could not be said of Dr. Robert Jeffress, pastor of my childhood church, First Baptist Church in Dallas, Texas, which enjoys a congregation of thirteen thousand members. Jeffress's sermons are broadcast on 11,295 cable and satellite networks in 195 countries. His radio show, *Pathway to Victory*, is distributed to more than nine hundred American stations, including major markets like New York City and Washington, D.C.

His website notes that he has made "more than 2000 guest appearances on various radio and television programs and regularly appears on major mainstream media outlets, such as Fox News Channel's *Fox and Friends*, *Hannity*, *Lou Dobbs Tonight*, *Varney & Co.*, and *Judge Jeanine*, also ABC's *Good Morning America*, and HBO's *Real Time with Bill Maher*."[8]

Jeffress also sits on President Trump's Evangelical Advisory Board, which was formed during the campaign in June 2016 to meet candidate Trump's "desire to have access to the wise counsel of such leaders as needed."[9] The advisory board, composed of an assortment of notable and influential evangelical leaders, was not disbanded after the election and has met with the president at the White House several times, even as recently as August 2018.[10]

It is indisputable that Jeffress possesses the evangelical gravitas that focus group member Mark Lee did not. But like Lee, Jeffress seems to believe that, when it comes to governance, Jesus needs to stay in His lane.

In an interview in 2016 with Mike Gallagher, a conservative radio talk show host, Jeffress described how he reacted to the question of whether he would prefer a president who governed according to the principles Jesus spoke of at the Sermon on the Mount.

"Heck no," Jeffress said. "I would run from that candidate as far as possible, because the Sermon on the Mount was not given as a governing principle for this nation."

He went on to say that governments are exempt from such biblical principles as forgiveness, or the willingness to turn the other cheek. "Government is to be a strongman to protect its citizens against evildoers," he claimed. "I don't care about that candidate's tone or vocabulary, I want the meanest, toughest, son of a you-know-what I can find, and I believe that's Biblical."

Jeffress held true to this interpretation of governance when offering support of President Trump's threats directed at North Korean dictator Kim Jong-un, in which the president warned, "North Korea best not make any more threats to the United States." He went on to say, "They will be met with fire and fury and frankly power, the likes of which this world has never seen before."[11]

In a phone interview with the *Washington Post* in August 2017, Jeffress said of Trump's remarks (which critics described as saber rattling) that "God has endowed rulers full power to use whatever means necessary," adding that this gives government "the authority to do whatever, whether it's assassination, capital punishment or evil punishment to quell the actions of evildoers like Kim Jong Un."

He went on to contend that Romans 12, which commands we "do not repay evil for evil," does not apply in the context of foreign policy, referencing again his belief that presidential decision making is biblically exempt from the principles laid out at the Sermon on the Mount. As he told the *Post*, "A Christian writer asked me, 'Don't you want the president to embody the Sermon on the Mount?' I said absolutely not."[12]

The Sermon on the Mount, which Jeffress is so quick to brush aside, is Jesus's most famous and cited sermon in the Gospels. It included guiding principles such as these: caution your tongue and the manner in which you present yourself (Matt. 5:33–37), do not seek vengeance (Matt. 5:38–42), don't be braggadocian (Matt. 6:1–18), don't follow the crowd (Matt. 7:13–14), and, importantly, be cautious about who you trust as your teachers (Matt. 7:15–23).

These, among others, are the principles that a major influential Christian evangelical leader who sits on the president's Evangelical Advisory Board says should be run from "as far as possible" when choosing a president.

And Jeffress wasn't simply saying he could look past someone not holding to the specific principles spoken at the Sermon on the Mount. Similar principles are not hard to find in good people of other faiths or of no religious faith at all. No, Jeffress was saying he prefers the *opposite*. He's saying that it is *good* in this context to be bad.

In essence, Jeffress was making the case that Donald Trump's sinful nature is a virtue.

This is actually much more antithetical to Christian teachings than focus group member Mark Lee's claim that he would check with Trump before believing Jesus about world affairs. Jeffress is essentially saying he wouldn't even ask because Jesus, apparently, wouldn't get it.

This is an example of the "wise counsel" that Trump sought at his Evangelical Advisory Board's inception.

Reading the words of such a prominent evangelical leader so flippantly abandoning the teachings of the Bible in favor of a political figure is a remarkable situation to witness, and it is an increasingly common occurrence.

This view of Trump as God's inerrant vessel permeates the modern evangelical movement, from its leadership down to its pews, and that's because it has a basis in a specific evangelical prophetic tradition.

Prophecy

Cyrus the Great was a Persian king and an important Old Testament figure who did not worship the God of Abraham but, according to both Christian and Jewish beliefs, was a vessel of God's will. His deeds were foretold by Isaiah, and his achievement changed the course of history for the Jewish people.

He was described through the prophet Isaiah as God's shepherd (Isaiah 44:28) and was "anointed" (Isaiah 45:1). It was said that God would use this man to "subdue nations" and that God would "open doors" for him (Isaiah 45:1), and that's exactly what happened.

Important to the story of Cyrus, it was foretold that he would accomplish all these things despite not knowing God. "I equip you," Isaiah 45:5 says, "though you do not know me."

In the time of Cyrus, around 586 B.C., Jerusalem had been conquered by the Babylonians, and God's temple there had been burned to the ground. The Jewish population had suffered either death or exile.

But Cyrus defeated the Babylonian army in 539 B.C. and, upon assuming the rule of Babylon, issued a decree that would bring displaced Judeans back to Jerusalem, as well as promising to rebuild

God's Temple, which the Babylonians had destroyed nearly fifty years prior.

The decree read, "Thus says Cyrus king of Persia, 'The Lord, the God of heaven, has given me all the kingdoms of the earth, and he has charged me to build him a house at Jerusalem, which is in Judah. Whoever is among you of all his people, may the Lord his God be with him. Let him go up.'" (2 Chronicles 36:22–23).

So in a nutshell, God used a great leader, one who did not know Him, to fulfill His purposes. Who Cyrus was, the nature of his character, whether he was a good person or a bad person, were irrelevant to the fulfillment of those purposes. Evangelicals have often used Cyrus's rise as an analogy to justify the rise of Donald Trump.

In 2018, when President Trump decided to move the American embassy in Israel to Jerusalem, a move celebrated by most on the American political right as a long-overdue gesture toward what many consider America's greatest ally, it was billed as a message that America recognized Jerusalem as the capital of Israel, despite the objections of the surrounding hostile nations and local Arabs. It was also a move that prompted a comparison of Trump to Cyrus the Great by the president of Israel, Benjamin Netanyahu.

During a visit to Washington, D.C., in March 2018, Netanyahu said of the move, "We remember the proclamation of the great King Cyrus the Great—Persian King. Twenty-five hundred years ago, he proclaimed that the Jewish exiles in Babylon can come back and rebuild our temple in Jerusalem." He added, "And we remember how a few weeks ago, President Donald J. Trump recognized Jerusalem as Israel's capital. Mr. President, this will be remembered by our people throughout the ages."[13]

Interestingly, while Cyrus's actions were enormously consequential to the history of the nation of Israel and the Jewish people, his actions were not especially unique for Persian conquest.

Whereas the Babylonians would scatter the populations of those they had conquered in order to lessen the opportunities for them to gather and form a rebellion, Persians would permit the conquered populations to continue their own religious practices and would even work with them to rebuild what had been destroyed during the conquest. The reason that's an important point is because these comparisons of Trump to Cyrus seem to miss the fact that these vessels are not themselves required or even encouraged to be revered.

Certainly there have been vessels of God's will that the Bible makes clear are worthy of admiration and reverence, but that is when, for instance, they are used specifically as a statement of something else, such as faith. Moses could easily be described as a vessel of God's will, but beyond the particular actions he took that advanced God's plans, he was also a person who, despite personal failings, grew in his faith, and the actions he took were directly tied to that faith.

In fact, it would not be hard to argue that a "vessel" is not even required to commit good acts. Judas committed acts of evil against Jesus, acts that Jesus Himself foretold at the Last Supper. His acts were undoubtedly part of the tapestry of events that led to the crucifixion, the single most consequential event in the entire Bible, yet I'm not sure that anyone would describe his actions as "good," much less his person.

Like Moses, part of the story of Judas involved his character and his faith, even as much as his actions, though in his case the opposite end of the spectrum from Moses. Yet he still fulfilled God's purposes as part of a larger plan. He was a vessel.

But this was not the case with Cyrus. The truth is we don't really know that much about the character of Cyrus because his character wasn't the point, nor was it even slightly relevant. He was the ruler of a particular land at a particular time whom God granted a path to victory in his conquests. Once that conquest had been completed, he exhibited typical Persian rule, which resulted in Jerusalem being home to the Judean people following decades of exile.

He had a purpose; he fulfilled that purpose.

So doesn't that prove that Christians are right to compare Cyrus to Trump? Given that their chief argument seems to be that his character doesn't matter in light of his works?

Well, there's a glaring difference.

Cyrus was named by God's prophets at the time as existing to fulfill a purpose, a purpose that was explicitly stated within the prophecy itself. Donald Trump's purpose? That was named by man. In other words, it is the difference between God declaring His will and man declaring God's will. One is faithful. The other? It is at least fraught with danger as is always the case when men attempt to speak of God's intent.

But don't worry. Trump evangelicals don't necessarily believe that they are declaring God's will. They know he is God's vessel for good ends because Trump has prophets of his own.

In 2018, David Brody, the chief political analyst for Pat Robertson's Christian Broadcasting Network, and Scott Lamb, vice president of Liberty University and a columnist at the *Washington Times*, wrote what they referred to as a spiritual biography of Trump called *The Faith of Donald J. Trump*.

Shortly after its release, Trump tweeted about it: " 'The Faith of Donald Trump,' a book just out by David Brody and Scott Lamb, is a very interesting read. Enjoy!"[14]

The book often goes back and forth between describing Trump's faith and purpose as unknowable and telling stories that seem to be designed to imply the divine purpose so many Trump evangelicals have attributed to him.

For instance, the authors provide the account of Lance Wallnau, a Dallas business owner who prophesied during the primaries in 2015 that Donald Trump would become president.

Wallnau tells of how he ended up attending a meeting at Trump Tower in the fall of that year after Kim Clement—a Pentecostal Christian who was scheduled to be at the meeting to deliver a message to Trump described as a "word from the Lord"—suffered a stroke that prevented his attendance and Wallnau was invited to take his place.

Upon returning home from the meeting, Wallnau explained to the authors, he still wasn't sold on Trump as a candidate he could endorse. But it was then that God spoke to him.

"The Word of the Lord that was going to come to Kim about the presidential candidate—it came to me," Wallnau said. "I heard these words: 'Donald Trump is a wrecking ball to the spirit of political correctness.'"

Apparently God had been reading Twitter and was as fed up as everyone else with such atrocities as saying "Happy Holidays" instead of "Merry Christmas."

But the prophecy wasn't complete.

Before taking a second trip to join a meeting with Trump, Wallnau, who says he was not one to receive direct messages from God prior to this, got another message. This time, the voice of the Lord said, "The next President of the United States will be the forty-fifth president, and he will be an Isaiah 45 president."

Not knowing what "Isaiah 45" meant in this context, Wallnau cracked open his Bible and discovered the story of none other than Cyrus the Great, a pagan leader who freed the people of Israel. Wallnau says the verse to which he was led reads, "Thus says the Lord to Cyrus whom I've anointed," which told him that Donald Trump was a modern-day Cyrus and that he, this "wrecking ball," was destined to be president.

This clinched it for Wallnau, as well as (one can safely assume, given its inclusion) the authors of *The Faith of Donald J. Trump*. From the book:

"That line freaked me," Wallnau said, because "I always teach that salvation's free but the anointing is given by God for certain tasks, and I always thought about it as a Christian currency, of being anointed or blessed. But in this case, it says God anointed a heathen ruler. And that was the moment I realized God was going to anoint someone who doesn't know him."[15]

At the meeting Wallnau had been scheduled to attend, he told Trump about his prophecy. After explaining the story of Cyrus, Wallnau said, "I told Trump that the reason God's hand was on him was because of the grace of God—to restrain evil. And that he would be the forty-fifth president of the United States. And when it was over, Mr. Trump knew that I had prophesied over him."

So there you have it. Trump is not simply comparable to Cyrus the Great, he is the prophesied reincarnation of the man.

But Wallnau was far from alone in viewing Trump's rise as confirming God's intent to use this president as a vessel for His good works. Or that Trump is a leader whose mission is anointed and whose existence is prophecy.

In October 2018, *The Trump Prophecy* was screened in more than a thousand theaters nationwide and told the "true" story of Mark Taylor, a retired firefighter who says that around two a.m. on April 28, 2011, he caught Donald Trump on C-SPAN and heard God say to him, "You are hearing the voice of the next president."

The film dramatizes Taylor's five-year journey of gathering others who claimed to have heard this same message as they prayed and urged God to elect Donald Trump, finally seeing their prayers answered in November 2016.

Taylor's 2017 book, *The Trump Prophecies: The Astonishing True Story of the Man Who Saw Tomorrow ... and What He Says Is Coming Next*, promises to show readers "how the Lord can and will use anyone regardless of education, background, or health to communicate His message to the Church of Christ." It refers to Trump's election as "one of the most incredible miracles our country has ever seen."

Eight hundred eighty-nine Amazon reviews later and the book still has a 4.5 out of a possible 5 stars, with comments from readers like "Lou," who said he was "thrilled to realize that it actually was God who orchestrated a scenario that would bring an imperfect but willing man into the White House under IMPOSSIBLE odds to show His glory and power and plans for America."

User "John and Tina" noted that this was a great book for those who still had lingering doubts about their choice of Trump over Hillary Clinton, saying, "Dislike Trump? Hate Hillary? Read this book. There is more going on than you think. 'We, the people' are not in control. God is."

This is great news for those Christians who struggled in 2016. With prophecy, bothersome concerns like character flaws and the spoiled fruit Trump bears on a regular basis are irrelevant! Because, like Cyrus, who he is matters less than what God has planned for him. And don't worry, this isn't men speaking on God's behalf; God literally told them this was the case.

Forget the condition of Trump's soul! Let's Make America Great Again!

Compartmentalization

Of course, this line of argument doesn't convince everyone. After all, it is quite a leap to look at serial adulterer, porn-magazine enthusiast Donald Trump and come up with "foretold deliverer of the church." Don't worry though—if you can't baptize Trump as a prophesied king, you can use the handy "separation of church and state" argument to write God out of the equation altogether. It's just a matter of compartmentalization.

The pillars of this new evangelical ethical hierarchy were being established long before the concluding events described in the previous chapter. They were only seeds at the time, but over the years of his presidency thus far, they have taken root as foundational to Trump evangelical doctrine.

Late in the election, after Falwell Jr. had become more established as a surrogate for candidate Trump on the television news circuit, Falwell Jr. took upon himself a task that I can describe only as "moral translator." It was not just him, but he's a perfect example of the practice: Christian religious leaders stepping before their flocks and explaining away aberrant or abhorrent behavior. To make that behavior palatable or passable. To translate bad past deeds into good future outcomes.

Falwell was definitely a trailblazer in this regard.

It was in this role that Falwell first took to the cable networks like a new pundit or, dare I say, prophet of Trump—a role he still enjoys to this day. In once such appearance, Falwell Jr. was on CNN to discuss the mounting accusations from several women who claimed that Donald Trump had sexually harassed and even assaulted some of them. Falwell's task was clear. He had to make this scandal a nonfactor for evangelicals.

Erin Burnett, host of *Erin Burnett OutFront*, asked rather directly whether or not Falwell Jr. would support Trump if he was proven to be a sexual abuser. Falwell Jr. hemmed and hawed and offered some boilerplate about everyone being redeemable. Using John 8:7, he offered this telling take on the situation: "I don't know who was a worse womanizer, I don't know who assaulted more women; John F. Kennedy or Bill Clinton. But I can tell you, John F. Kennedy did the right things. He cut taxes, he brought prosperity, and I would vote for John F. Kennedy again if he were on the ballot today because of his conservative leadership."

It's a telling answer, in that Falwell is essentially acknowledging that the "but abortion" rationalization that is often tossed around would not stop Falwell from voting for a Democrat if that candidate was more likely to cut taxes. But we'll get to that point later. It was the next part that was especially revealing: "We're not electing a pastor," he said, "we're electing a president."

And there you have it. That's it, you see. That was the new truth, and he was telling it on the mountain. It's okay, says the new truth. It's okay to vote for the womanizer, the abuser. It's fine. As long as they cut taxes. Godliness is neato, but self-interest is in the driver's seat when it comes to following a national leader.

As crazy and overt as this was, he wasn't finished.

"Our country is going to suffer if we get sidetracked on these rabbit trails about 'is this person a good person, is that person a good person,' it's not about that," said the offspring of the founder of the Moral Majority. "It's about 'what are their positions on the issues?'" he added.

Jerry Falwell Jr., the head of the university founded to pass on Christian values to its students, to pass them on *specifically so they can spread out into the population and culture*, was emphatically and explicitly stating that it is not the responsibility, or even a matter of any particular interest, for Christians to support or follow or vote for someone who shares those values, or at least does not openly defy them.

What Falwell Jr. was preaching states that it is the calling of the devout to put aside glaring assaults and affronts to morality in favor of winning a political race.

This idea that you bring God only into situations that "pertain to Him" ran counter to just about every prior teaching of Christian faith as a part of American life that I'd ever been taught. It feels new, in that Falwell Jr. and those like him are expressing that this is not "making the best of a bad situation." He's suggesting that one does it by design, that it is preferable, that it is an *expression* of faith instead of a lapse of it.

Truthfully, it's not entirely illogical to believe that a Christian in America must choose a leader based on the practicalities of everyday life and the mundane necessities of a nation. After all, a person can believe there is a right and wrong way to do things that are not heavenly by nature.

I don't look to the Bible to figure out whether the toilet paper should drape over or under the roll, I just decide which is best. Why wouldn't a voter do so in an election, Christian or not?

Aren't goals like giving our preferred party the opportunity to govern or securing our national prosperity or sovereignty things that might override more esoteric concerns about the particular flawed human being elected to achieve them?

And, of course, this is exactly the argument people make.

"So what if he's a philanderer," they'd say to me. "Solomon had one thousand wives and concubines and God still used him for good!"

In other words, he's the vessel. And we do not question the vessel. What this line of argument does is claim that politics is a morally neutral sphere, like a supermarket, where choices between candidates have no more moral dimension than a decision between apples and oranges.

What has been surprising, though less and less so as time goes on, is that so many Christians accept—embrace, actually—the idea that there are areas of life where the Bible's moral standard simply "doesn't apply." There is this underlying belief that the biblical simply isn't applicable in day-to-day concerns. As though it is nothing more than a self-help book that is to be consulted when that makes sense and ignored when it's advantageous to do so.

I may be flawed and fallen away from godliness, but I don't share this view, and before all this madness unfolded, I still believed such an idea wasn't even close to a "mainstream" evangelical philosophy. I was certain many or even most people could easily slip into doing it, just as I was certain that competent Christian leaders and pastors would encourage them not to.

Still, and it bears repeating, I was wrong. Rejecting biblical standards of character for leadership has become the preferred point of view.

Dr. Robert Jeffress, a full five years prior to his declarations that we should "run like hell" from candidates who embody the principles of the Sermon on the Mount, was already professing this compartmentalization when making a political calculus.

In 2011, Jeffress made waves while introducing Governor Rick Perry of Texas at the Values Voter Summit in Washington, D.C. Jeffress had endorsed Perry in the Republican presidential primaries and of Perry's opponent, Governor Mitt Romney, Jeffress said, "I just do not believe that we as conservative Christians can expect him to stand strong for the issues that are important to us."

He referred to Romney's Mormonism as a "cult" and said he did not believe Romney was a Christian, then went on to say that Romney was a "conservative out of convenience" and "does not have a consistent track record on the subject of marriage, on the sanctity of life."

Once Romney secured the Republican nomination, Jeffress

changed his tune slightly, saying that while his concerns over Romney's religious beliefs were still a problem, they were "not the most prominent issue, even for Evangelicals," adding, "the economy and the social issues trump theology."

Not exactly a ringing endorsement, but it was the type of tacit support one might expect from a conservative Christian when left with only a few options. A "lesser-evil" choice, one might argue.

Yet, in this reversal you saw more of this compartmentalization that Falwell Jr. presented on CNN in defense of Trump. Theology and practicality were placed into different boxes and, alarmingly, practicality outweighed any biblical considerations.

Jeffress endorsed Romney over his own (I would argue misguided) theological objections because, as he explicitly stated, they were "not the most prominent issue."

He proposed, albeit with little enthusiasm, that Christians must, when it comes to something as monumental as presidential elections, accept what he'd previously denounced as unacceptable and let the chips fall where they may. After all, there might be tax cuts, and that, one would assume, trumps theology.

To be clear, my citation of this theological flip-flopping as an example of backward Christian priorities is in no sense a concurrence with Jeffress's claim that Romney's Mormonism was a "problem" for American Christians. I didn't then nor do I now believe it was or should have been.

Mitt Romney's good character had been sufficiently consistent throughout his life that I and other Christians who supported him could credibly claim—regardless of his personal religious beliefs— that his principles were in line with our own and, contrary to Trump's, not in grotesque open defiance to our stated values.

Most of my concerns about Mitt Romney centered around practical considerations related to his commitment to conservative political ideals. Having rested the question of whether or not his character was unfit for the office, and recognizing that this was decidedly not the case, I could then make my decision based solely on practical considerations. In other words, there was a *prioritization* to my decision making, not a *compartmentalization*. I considered the eternal first, then the short-term or long-term objectives a president would be tasked with achieving.

Jeffress displayed an entirely different calculus. Prior to Romney's nomination, he believed Romney's existence was antithetical to the principles we should expect from a leader that Christian voters could pull the lever for.

If he was sincere in this concern, it is an especially odd thing for him to claim that that concern *doesn't matter* when balanced against policy objectives and political expediency.

Again, that's not prioritization, that's compartmentalization.

In a 2016 interview moderated by pollster Frank Luntz, then candidate Donald Trump perfectly described the compartmentalization of spiritual and practical concerns, a necessary ingredient to adopt the version of Christianity that Trump evangelicals were selling.

As I wrote earlier, Trump was once asked about forgiveness, specifically whether it is something he ever sought from above. Trump replied, "I am not sure I have. I just go on and try to do a better job from there. I don't think so." In case it isn't clear, that is *not* the correct answer. Not for the devout. Or even the backsliding casual.

Trump then further revealed some confusion about the very concept of grace, saying, "I think if I do something wrong, I think, I just try and make it right." Again, incorrect.

Now, certainly this might be looked past in someone whose faith is fairly new, as we are sometimes to believe is true of Trump, but he went on to say something that defines exactly, perfectly, and succinctly what I've been talking about: "I don't bring God into that picture. I don't."

That is what he said. Just as the photo Falwell Jr. tweeted perfectly captured the new version of the Trump evangelical taking center stage, so Trump's brief answer impeccably encapsulates the compartmentalization that the individual Christian was undertaking to be a member of that movement.

Trump evangelicals are tasked with putting God in one compartment and their voting and political conscience in another. This is clearly out of step with Christian tradition, which has historically valued Christians who used their faith to influence public affairs. William Wilberforce's faith inspired his efforts to fight for the abolition of slavery in the British Empire. Christians like Sophie Scholl and Dietrich Bonhoeffer (to name only a couple) were political dissidents during the Third Reich.

Jesus said, "Render under Caesar what is Caesar's," referring to coinage bearing the image of Caesar. But he added, "and to God what is God's." Human beings bear the image of God. Politics involves much of "what is Caesar's" and much of "what is God's." It isn't a nonspiritual realm, and compartmentalization ignores this.

One often-overlooked impact of compartmentalization is the injury it does to the soul of Donald Trump. In *The Faith of Donald J. Trump*, Brody and Lamb certainly might have addressed the encounter between Luntz and Trump in a number of ways. The way they chose to do so exemplifies the doctrine of protecting Trump from consequence, even if doing so is a disservice to his personal relationship with God.

Rather than acknowledging the profession of a Christian faith that excludes the foundational premise of repentance, when Trump said he didn't think he'd ever asked for forgiveness, they offered spin, saying, "Ironically, if the truth of the matter is that Trump has never asked God for forgiveness, then if Trump had lied and said 'yes'—he could have avoided criticism. That is, lying about his practice of confession would have kept him out of trouble with the piety-inspectors."

In other words, the problem isn't Trump's fundamental misunderstanding of the foundational philosophy of Christianity: the idea that repentance is the key to eternal salvation. No, the problem is that those of us who might point out that his take is incorrect are "piety inspectors" for doing so.

The implication is that it is somehow "our" fault that Trump would *be forced* to lie to avoid our picky criticisms. Somehow Trump's error in seeing no reason to ask God's forgiveness is on *us* rather than on *him*. And to reiterate, that picky criticism of the "piety inspectors" is about the single most vitally important aspect of the Christian faith. It's the one thing that you would think these concerned evangelical Christians surrounding Trump would want him to understand, if merely for the sake of his own soul.

The authors of the *Faith of Donald J. Trump* were far from alone in contorting themselves to protect the doctrine of compartmentalization.

Compartmentalization has endless application and is, once undertaken, a virtually impenetrable barrier to conscientious decision making.

But when that compartmentalization is still not enough, and if prophecy and vessel theology aren't doing the trick either, there is one tried-and-true rationalization that is the final pillar of this brave new truth in service of Donald Trump. If you can't convince people that Trump is a new King Cyrus, God's gift to the GOP, and if they don't buy the compartmentalization argument that there is no spiritual dimension to political decisions, there is one more handy fallacy you can use to drive parishioners to the polls: the lesser-of-two-evils principle.

It's by far the thorniest argument to refute, because it's the one that has an element of truth.

Lesser Evils

Even after the surprise leak of the infamous "grab 'em by the pussy" audio from a 2005 *Access Hollywood* tape, evangelical leaders without the least bit of shame or even hesitation continued to rally the troops in support of Trump.

It was really something, actually, and goes to show how things have worked ever since. After that particular scandal hit, Gary Bauer, chairman of the Campaign for Working Families, said in reaction to the tape, "Hillary Clinton is committed to enacting policies that will erode religious liberty, promote abortion, make our country less safe, and leave our borders unprotected."

How very neat and tidy is that? *Poof*—any momentary self-doubt disappears in the glaring twin fires of fear of subjugation and lust for power.

Likewise, Steve Scheffler, head of the Iowa Faith & Freedom Coalition, intoned, "The Bible tells me that we are all sinners saved by grace, and I don't think there's probably a person alive that I know of that hasn't made some mistakes in the past."

Poof again. No need to worry about this—everyone is equally sinful, except for Democrats, who are worse.

Wayne Grudem, research professor of theology and biblical studies at Phoenix Seminary in Arizona, had penned an early "moral case" for Trump in July 2016, which he retracted in light of the tape, writing a new column in which he said he was no longer sure whom to vote for, given the situation, while at the same time urging Trump to withdraw from the race.

But by November, that column was gone, too. With weeks left before the election, Grudem put his concerns aside and articulated a common view of the conundrum "principled" evangelicals faced on Election Day.

Grudem outlined the race as starkly between two choices, Trump and Clinton, the infamous "binary choice" that was the subject of so many articles and debates that year. He argued that a third party was nonsense and "fantasy" and that voting third party was the same as voting for Clinton. He said that voting for Clinton was voting for her anti-Christian policy.

His conclusion was an obvious fait accompli:

It isn't even close. I overwhelmingly support Trump's policies and believe that Clinton's policies will seriously damage the nation, perhaps forever. On the Supreme Court, abortion, religious liberty, sexual orientation regulations, taxes, economic growth, the minimum wage, school choice, Obamacare, protection from terrorists, immigration, the military, energy, and safety in our cities, I think Trump is far better than Clinton (see below for details). Again and again, Trump supports the policies I advocated in my 2010 book Politics According to the Bible.

The basic premise of Grudem's eleventh-hour appeal was an old one and commonly employed during elections: choosing the lesser of two evils.

As November approached, support for this premise was widespread among evangelical leaders, and many made public appeals for voters to support it as well.

So we were faced again with the bottom of the rationalization barrel: "lesser evils."

But the fact is that Trump supporters apply moral hierarchicalism—the process of organizing moral decisions into a hierarchy, placing certain moral decisions higher than others when there is a conflict—as a go-to answer for just about every criticism of the president that can't be ethically defended on its own.

It is the source of more "whataboutism" arguments than just about any other subject under discussion in this era. Believe Trump's plans with tariffs are bad for the economy? Well, it's better than the socialism Clinton would've started steamrolling us with. Find Trump's reference to a poor country as a "shithole" offensive? Well, would you rather have a president who says rude things or a president who makes deals with Iran?

This (a) versus (b) mentality has become so overused in defense of Trump that it has spilled into everything else. The feeling at times is that partisans can get on board with just about anything, so long as they can find a way to claim that a greater good is being served.

And that is when it starts to get blurry. Is a hierarchy being employed to determine that there is an urgent moral dilemma? Or are people just employing moral relativism in order to feel okay about supporting actions they would otherwise find objectionable? That a debate on the right even exists over when and where these types of morally relativistic comparisons are appropriate or not appropriate is itself a departure from traditional conservative ideology. Moral absolutism has, in my experience, always been a pillar of conservative philosophy.

Conservatives, and certainly Christians, have long abhorred moral relativism, but in the age of Trump, it is starting to appear that those long-standing objections did not depend on good behavior so much as correct policy. George W. Bush or Bob Dole or John Mc-Cain all espoused popular conservative policies, but they also generally conducted themselves in such a way that defending them didn't require a declaration of a moral imperative.

They made mistakes, as anyone will. But Trump is and always has been something less than morally upright. His entire *brand* is to be morally indefensible.

And so, given this platonic example of a moral deviant, the GOP held its nose and said, "Well, he's the lesser of two evils."

Everything boils down to "lesser evil," and relativism is chic in the Trump GOP.

The Internet exploded in early 2019 when the recently elected senator from Utah, Mitt Romney, wrote an op-ed for the *Washington Post*, in which he criticized the public character of Donald Trump and offered concern that there is a national and cultural cost to excusing it, as well as an impact to our standing in the world.[16]

"Trump's words and actions have caused dismay around the world," Romney contended, citing a Pew Research Center poll that saw residents of Germany, Britain, France, Canada, and Sweden as having lost faith that America's president will do the right thing in world affairs, with only 16 percent saying yes, down from 84 percent just one year prior.

The question of Trump's character, as well as its impact on the country and the world, is certainly a debatable topic, and reasonable people may disagree.

But among Trump supporters, it is irrelevant for the simple fact that the alternative to Trump would have been Hillary Clinton, who, in their view, is not merely objectively worse but so bad that we have a moral obligation to look past Trump's weaknesses given the danger she, and any Democrat, poses.

That seems like a bit much.

Roger Kimball, while not an evangelical to my knowledge, is a conservative commentator who wrote a rebuke to Romney's op-ed in which he highlighted the many policy achievements of Trump (which, you may notice, is not really a rebuke of Romney's criticism). While he didn't explicitly say so in the op-ed piece, his implication was that these character defects, whatever they are, fall short when balanced against the objectives of policy.

But I'm not merely guessing about what was left unsaid. Just a few days prior, in an article in *American Greatness*, Kimball had made clear where he stood on the question of character when he challenged the notion that such a thing is even quantifiable outside of comparison to other concerns. "Let us grant that the president is an imperfect man," he wrote. "What betokens worse character: tweeting rude things or having sex with your intern in the Oval Office? What's worse, insulting Bob Corker or using the Department of Justice and the IRS to harass and persecute your political opponents?"

This is the heart of moral relativism, and though Kimball may not represent the Trump evangelicals I've been describing in these pages, his rationale perfectly encapsulates their "lesser evil" argument.

While vessel theology and King Cyrus eschatology require one to accept certain supernatural arguments that many find outlandish, and compartmentalization means defying years of Christian teaching about political engagement, the "lesser evils" principle is a great wedge to force reluctant Christians into the pro-Trump camp. It doesn't require the existence of a good, prophetically approved option, nor does it do the opposite by assuming that there is no place for spiritually informed decisions in politics. Instead, it says that when we're faced with bad options, we make a utilitarian choice and vote against the worst option.

This argument is the most effective because, in fact, this is a form of prioritization. We do have to make choices between bad options all the time. That's simply a consequence of living in a fallen world. So it's not an inherently bad argument, but it's been employed hypocritically. We'll get into why that is later in the book, but for now, it's enough to explain its role as a key doctrine of Trump evangelicalism.

With all of the needed rationales in place, some two years after the election, for many, the Good News in Christendom—that the kingdom of God is coming—has gotten an update. The *New* Good News is that Donald Trump is making America great again.

To recap how the New Good News works:

- *Vessel Theology*: God used the harlots and He used Cyrus to achieve His good ends, and like them He can use Trump to fulfill the evangelical policy agenda, which is itself a godly agenda and thus a good end. Ergo, do not question the vessel, or you are questioning the limits of God's influence. God can use anyone! (Except Hillary Clinton.)
- *The King Cyrus Argument*: Do not worry about whether or not He *can* use Trump this way. He *will* use Trump this way. It has, after all, been prophesied.
- *Compartmentalization*: But even if you don't believe those prophecies, and you still find yourself concerned that Trump seems at times to be untrustworthy or of low personal character, have no fear! All those principles Jesus spoke of at the Sermon on the Mount are actually things we shouldn't want in a leader; in fact, we should want the *opposite* in a leader! Literally, being like Jesus would be super-bad in this scenario.
- *Lesser of Two Evils*: Still not convinced? No problem. There's always the "lesser evil" argument to consider. And given that

anyone who is running against the Republican candidate is literally Satan, the moral calculus is always easy.

Trump's unapologetic and arguably narcissistic approach to politics undoubtedly helped him secure the nomination against the backdrop of more traditional candidates, who were seen as spineless and pandering. He was viewed as speaking the truth, no matter how uncomfortable, and whenever that truth telling seemed more like cruelty, he dismissed those concerns as weak.

Evangelicals in recent years have taken it upon themselves to adopt those same attitudes in defense of anything Trump says or does. This goes way beyond typical political spin and water carrying. They seem to believe it is their job to protect this president, to shield him. Even from God.

When confronted with an indisputably sinful act he committed in the past, they accept Trump's apologies. Apologies he never offered. They talk about spiritual growth and the importance of redemption. Growth he's rarely exhibited and redemption he's never sought.

You do not question the vessel.

A difficult truth to contend with in the age of Trump is that perhaps the stigma I described in chapter 1, the caricature of the evangelical, was appropriately placed. Or at least not entirely *inappropriately* placed. That is, perhaps the stereotype has a basis beyond Donald Trump.

Of course, it would be foolish merely to claim someone is a hypocrite or dishonest without demonstrating that he held any defensible views to betray in the first place. Unless you have a moral code to fail to live up to, you cannot be accused of failing to live up to such a code.

So far I've discussed how the movement embraced secularism

through Trump, and I've discussed how in his presidency they have claimed that, by supporting a movement that is often displaying an open defiance to biblical values, they are simultaneously protecting those same values.

So if this is the New Good News, as I have said, then what was the Old Good News?

Chapter 3

THE OLD GOOD NEWS

T hey are very dangerous. They are not differentiating between Caesar and God. They are trying to equate Christianity with Americanism."

Anyone who has been keeping up with the American political landscape these last few years would not be surprised to read the above quote and discover it is a criticism of the American evangelical movement. What may surprise you is when this criticism was leveled.

1980.

The quote comes from Dr. George Jones, director of the Northern Kentucky Baptist Association. Jones, a minister and a Republican since 1936, was speaking to the *Washington Post* about the rise of political evangelicalism with organizations like Jerry Falwell's Moral Majority leading the way and touting some 72,000 pastors among its members.[1]

In the same *Washington Post* article, Falwell says of then-candidate Ronald Reagan that they will mobilize for him "even if he has the devil running with him, and we'll pray he outlives him."

Despite also being an evangelical, Jones vehemently disagreed with the evangelical movement's quickly formalizing relationship with politics. "They're one-issue people," he said. "We must keep church and state separated."

"They won't last long," he predicted.

If you're reading this book, you know his prediction was wrong. The modern evangelical movement could be described as more powerful than ever.

But is it the same movement?

The roots of American evangelicalism can be traced back hundreds of years, with much of the twentieth century up through the 1970s seeing the movement attempt to grow influence over national policy, with some notable successes such as their involvement in the passage of prohibition.

In *The End of White Christian America,* author Robert P. Jones discusses how on the heels of the "success" of prohibition followed a drive among Protestants to establish a "firm foothold in the heart of American political life."[2] Fund-raisers were held to finance new buildings in Washington that expressed the goal of establishing "a Protestant presence of Capitol Hill," which Jones summarized as a "platform for stamping federal legislation with Protestant morality, for leveraging the power of politics to usher in the Kingdom of God on earth."

That idea—that America's role in the world, and in the spreading of God's kingdom, is unique, vital even—is one that continued to drive the growth of evangelicalism as a political force, eventually culminating in the birth of what is described as the modern evangelical movement.

President Jimmy Carter in many ways is a prominent feature of what helped the movement find more solid footing in national politics, which had been sought for so many decades prior.

Falwell organized a series of rallies called "I Love America"; these rallies sought to link political agendas directly to the faith of evangelical Christians. Falwell then leveraged the success of these rallies, along with his increasing prominence as a vocal critic of the Carter administration, to found the Moral Majority in 1979.

The organization would find success almost immediately in helping secure the election of President Ronald Reagan the following year, after running more than $10 million in ads.[3]

If today's evangelical movement could be described as married to the Republican Party, this victory in 1980 by the Moral Majority and other prominent evangelical figures and organizations was almost certainly the wedding.

So successful was this union that by 1989 the Moral Majority had disbanded, having declared its mission accomplished.

It is hard to argue that they were wrong.

The pervasiveness of the evangelical movement and its hold over the language and rhetoric of the Republican Party would become most evident in the 1990s in the fight against Bill Clinton.

A Question of Character

On August 17, 1998, an embattled President Bill Clinton took to the airwaves to admit to something he'd been denying for some time. After mounting pressure, and in the wake of his testimony in an ongoing investigation that would reveal the truth and its stark contrast to his denials, President Clinton finally admitted on national television to having had an affair with White House intern Monica Lewinsky.

"Indeed, I did have a relationship with Miss Lewinsky that was not appropriate. In fact, it was wrong. It constituted a critical lapse in judgment and a personal failure on my part for which I am solely and completely responsible."[4]

The entire ordeal was a watershed moment in what had become known as the "culture war," the catch-all name for events that pit traditional and progressive values against each other and that are often played out in the public spotlight of Hollywood and Washington.

At the time, I was a line cook at a national chain restaurant where most conversations did not center around politics. But there was no escaping the conversation that these events encouraged between Americans of all political stripes and class distinctions.

At some point during the madness, a coworker asked for my opinion, and I said something along the lines of expecting more from our leaders than what we were getting. I recall very well her frustration as she questioned that mentality, expressing first that the president's private affairs were "none of our business," and then going on to point out the high likelihood that other presidents, such as JFK, had allegedly engaged in the same types of behaviors.

"He just got caught" was her summation. And she did not believe that merely being the one who had been caught among a sea of hypocrites was worthy of all the attention and the talks of impeachment, and it certainly didn't undo the other aspects of his presidency, such as a booming economy and relative peace with other nations.

I was only twenty-one years old at the time, but, having grown up with a conservative Christian foundation, I experienced an immediate distaste for what I would eventually come to know as "whataboutism": the tactic of drawing attention away from one wrongdoing by pointing to another.

In the case of my coworker, it was an especially meaningless gesture, since while she said the behavior didn't matter, she simultaneously showed disdain for others who had committed the same behavior. It seemed to me that if it's bad, it's bad, and pointing to others would not change that.

Upon addressing that point, she explained that we aren't all Christians and we don't all share the same values. She then proceeded to use the same poorly interpreted Bible verse we've all heard a million times: "Judge not lest ye be judged."

But while this may have been the position of non-Christians, the overwhelming position of evangelicals concurred with my assessment at the time.

Ralph Reed, cofounder of the Christian Coalition, speaking to the *New York Times* in 1998, said of the Clinton affair, "Character matters." He added, "We care about the conduct of our leaders, and we will not rest until we have leaders of good moral character."

As reported at the *Washington Post* in September 1998, Christian televangelist Pat Robertson, addressing members of the Christian Coalition he'd cofounded with Reed, said, "President Clinton has so 'debauched, debased and defamed' the presidency that resignation is too easy an out."[5]

Of Clinton's television admission of wrongdoing, Jerry Falwell Sr. commented, "I did not see him look into the camera and say, 'I was wrong and I ask for forgiveness,' and I did not hear him take the burden of responsibility for his staff who stood with him."

And though Falwell believed forgiveness should be granted, his position on consequence was that Clinton was "no longer worthy to fulfill this office."[6]

The reasoning among evangelicals for why President Clinton should have stepped down in the wake of his admission was almost without exception a rallying cry for expectations of decency and character from our elected officials. Essentially, if the president of the United States had shown poor character to this degree, he must—in the best interests of the nation's citizens who may look to him for leadership—resign or risk a further decline of the values many of these Christian leaders believed were foundational to America's success.

He was immoral and a person of low character; therefore, he couldn't lead. Or shouldn't. That was the argument.

In today's climate, and given what we went over in the preceding chapters, the contrast is stunning.

Dr. James Dobson wrote a letter to the large membership of Focus on the Family that offers greater insight into this rationale:

> *As it turns out, character DOES matter. You can't run a family, let alone a country, without it. How foolish to believe that a person who lacks honesty and moral integrity is qualified to lead a nation and the world! Nevertheless, our people continue to say that the President is doing a good job even if they don't respect him personally. Those two positions are fundamentally incompatible. In the Book of James the question is posed, "Can both fresh water and salt water flow from the same spring" (James 3:11 NIV). The answer is no.*[7]

The verse Dr. Dobson quotes is on point and, in full context, further illuminates the importance of moral leadership.

JAMES 3:1–12 (NIV)

1 Not many of you should become teachers, my fellow believers, because you know that we who teach will be judged more strictly. ² We all stumble in many ways. Anyone who is never at fault in what they say is perfect, able to keep their whole body in check.

3 When we put bits into the mouths of horses to make them obey us, we can turn the whole animal. ⁴ Or take ships as an example. Although they are so large and are driven by strong winds, they are steered by a very small rudder wherever the pilot wants to go.

5 Likewise, the tongue is a small part of the body, but it makes great boasts. Consider what a great forest is set on fire by a small spark. ⁶ The tongue also is a fire, a world of evil among the parts of the body. It corrupts the whole body, sets the whole course of one's life on fire, and is itself set on fire by hell.

7 All kinds of animals, birds, reptiles and sea creatures are being tamed and have been tamed by mankind, 8 but no human being can tame the tongue. It is a restless evil, full of deadly poison.

9 With the tongue we praise our Lord and Father, and with it we curse human beings, who have been made in God's likeness. 10 Out of the same mouth come praise and cursing. My brothers and sisters, this should not be. 11 Can both fresh water and salt water flow from the same spring? 12 My brothers and sisters, can a fig tree bear olives, or a grapevine bear figs? Neither can a salt spring produce fresh water.

In this chapter, James goes to great lengths to warn those who would lead and teach others. A burden is implied. The burden of expectation, that those who might stumble are more likely to do so in service of someone they believe is their mentor.

A leader whose judgment cannot be trusted or whose vices run contrary to his stated values will either be seen as a fool or a hypocrite. And as Dr. Dobson points out, once sin is permitted in the name of achieving other ends, the perilous journey of lowering expectations can lead to the downfall of a people who no longer seek wisdom or morality, but instead worship at the altar of self-interest.

And yet, self-interest is a powerful incentive. For anyone hoping to remain principled, and both left and right claim that they do, self-interest can sometimes clash with those principles. But in these situations, intellectual dishonesty is the key to self-delusion.

The 1990s were a time of growth and peace for America. Historians can and will debate how much of that was a result of President Clinton, but the fact remains that while president, he oversaw a prosperous nation.

Many of the rationalizations in defense of Clinton centered around the simple notion that a man's personal failures are not inherently reflective of his capacity for greatness, nor were they a valid critique of his effectiveness.

Choice examples of this defense of Clinton were laid out and decried in the Dobson letter:[8]

> Let me share just a few of the hundreds of statements, in print and in the media, that exist on the record. You'll quickly recognize this effort by the press to undermine the moral values that we called "character." Hold on to your hat.
>
> "... we can remember that we are electing not clergy but political leaders—who need to be principled and devious, compassionate and brutal, visionary and, sometimes, utterly egotistical. If we try to do much better, we will end up doing worse."
> —SUZANNE GARMENT, SAN DIEGO UNION-TRIBUNE, 1992

> [Speaking on behalf of New York University media scholar Jay Rosen], "There is an important distinction between public and private character. **What candidates do in private is largely irrelevant,** says Rosen. What matters is their public conduct."
> —JEREMY IGGERS IN THE MINNEAPOLIS STAR-TRIBUNE, 1992

> "He [Clinton] will shave, wheedle, compromise and cajole until he finds—or creates—common ground. He is notorious for his ability to impress strangers and disarm opponents. He is notorious for leading people to believe that he agrees with them entirely ... without ever committing himself to their position. **This is a gift given only to the best politicians. It is how difficult things get done.**"
> —JOE KLEIN, NEWSWEEK, 1994

> "Whether character is a factor or not is relevant only as it relates to what the people want in terms of a President. They're

looking for someone with the character to get the economy back on track and answer the more serious questions facing this country."

—MAX PARKER, A CLINTON SPOKESWOMAN DURING THE 1992 PRESIDENTIAL CAMPAIGN

"Voters re-elected Clinton despite widespread doubts about his character. In CNN's election day exit poll, most voters continued to say Clinton is not honest and trustworthy. They've re-elected him because of his job performance—and crossed their fingers that character would not prove to be a major problem."

—BILL SCHNEIDER, CNN, 1996

"He has vacillated on issues large and small, and at times he has conducted himself like a man with something to hide. Nevertheless, we think he is still a better choice . . ."

—ST. LOUIS POST-DISPATCH, 1996

". . . Clinton was able to defuse the 'character' issue by focusing on voters' own wants and needs. They put their own interests above that issue, and thus relegated all the stories about Clinton's character to the back burner, or to the trash can . . . **it means that women and families have decided that it's more important to have their own issues addressed rather than worry about the character issue."**

—ROBERT A. JORDAN OF THE *BOSTON GLOBE*, 1996

Clinton is not the only politician in either party who lacks character, certainly, but he is the only one in American history, to my knowledge, who has been specifically applauded for his deceit. Let me share one of the most graphic illustrations of that support.

Please read carefully the following statement by noted syndicated columnist Richard Cohen, after Clinton's first term.

> "... he [Clinton] has been accused of adultery, sexual harassment, and ducking the draft—allegations that send some people into a frenzy of Clinton-hating. The President's ultimate sin, it seems to some people, is that he appears to have broken the rules—and gotten away with it. That is unforgivable. But to the rest of us, the character issue just hasn't taken. If we have learned anything over the last four years, it is that strictly personal behavior—in other words, sex—might be interesting, might be titillating, and might be even downright riveting ... One can argue that in both his triumphs and his failures there is a connection between the private and public Bill Clinton. **But once the public man is known, the private one just doesn't seem to matter anymore.... In his own way, Clinton taught us all a lesson about personal character that we should all remember the next time around: It's sometimes more interesting than important."**
> —RICHARD COHEN OF THE *WASHINGTON POST*, 1996

In all of these incidents that Dobson cites, the theme is the same. There is an acknowledgment of immorality and lack of character, but other things are deemed more important, such as the economy.

However, Dobson does not necessarily imply that these individuals are breaking their own moral code to rationalize their support of Clinton. Instead, he seems to be indicating that whatever the personal views of those quoted, the positions they have taken are in stark contrast to what Christ asks of His followers, and that widespread adoption of those views, even among non-Christians, in what he believes is ostensibly a Christian nation, would be a contributing factor to the moral decline of American values.

Dr. Dobson further addressed the fallacy that "the ends can justify the means" in his letter to his audience, specifically identifying it as a poor basis for determining what is worthy of endorsement.

"Because the economy is strong, millions of people have said infidelity in the Oval Office is just a private affair—something between himself and Hillary. We heard it time and again during those months: 'As long as Mr. Clinton is doing a good job, it's nobody's business what he does with his personal life.'

"That disregard for morality is profoundly disturbing to me. Although sexual affairs have occurred often in high places, the public has never approved of such misconduct. But today, the rules by which behavior is governed appear to have been rewritten specifically for Mr. Clinton."

Dobson clarifies that Clinton's behavior could impact children—especially girls, who he believed may interpret political rationalizations as endorsements—and give them the impression that such behavior is to be expected from men.

At any given time, 40 percent of the nation's children list the President of the United States as the person they most admire. What are they learning from Mr. Clinton? What have we taught our boys about respecting women? What have our little girls learned about men? How can we estimate the impact of this scandal on future generations? How in the world can 7 out of 10 Americans continue to say that nothing matters except a robust economy?

As George Washington said in his presidential farewell address, "Of all the disposition and habits which lead to political prosperity, Religion and morality are indispensable supports . . . And let us with caution indulge the supposition, that morality can be

maintained without religion . . . reason and experience both forbid
us to expect that national morality can prevail in exclusion of
religious principle."[9]

The assertion, today supported even by some Democrats, is that
when all consideration is weighted in favor of the economy or a
president's agenda, even and in spite of the lowering of expectations for decency, the cost is too great.

Let's again break that down into a simple form: some things are
just more important than others.

In closing the letter, Dobson declares that "nothing short of a
spiritual renewal will save us."

Again, what Dobson goes to great lengths to impress upon his
readers is that the ends cannot justify the means. Morality cannot
be cordoned off from the public square.

It is Dobson's position, and I would argue the biblical position,
that a justification of any means in the pursuit of ends is explicitly
warned against in the Bible, and that as Christians, we should be
especially concerned about an argument that seems to imply self-
interested motivations, such as the economy, are of a higher priority than the character of our leaders.

In summary, Ralph Reed, James Dobson, Pat Robertson, Jerry
Falwell Sr., and scores of other evangelicals took the position in
light of the increasing evidence for a lack of moral character on the
part of Bill Clinton that he should resign the presidency for the
good of a nation already desensitized to moral depravity.

They believed that there would be a long-term cultural consequence if he were to stay in office, and that the decline of expectations
of character or the acceptability of adultery and hedonism, already
something they viewed as a growing problem in the culture, would
steepen the downward trajectory of American values.

And perhaps most important, the evangelicals went to great lengths to offer warnings that the ends cannot justify the means, and that to believe they can is to head down a perilous path that would lead to unintended bad consequences. Thomas More, in Robert Bolt's play *A Man for All Seasons*, sums up their position well: "When statesmen forsake their own private conscience for the sake of their public duties, they lead their country by a short route to chaos."

In a general sense, the criticisms they offered were often designed to be acceptable to people outside their faith as an appeal to simple decency. But it was also very clearly a warning to their flocks. A warning that cautioned those who might choose self-interests such as the economy or relative peace and security as a substitute for adherence to God's morality.

While much of what they warned might happen in the wake of not "doing the right thing" concerned God's view of things, it was clear they were arguing that these types of decisions can have long-lasting cultural consequences.

Broken Compass

I will not pretend that these leaders I've referenced were motivated by their desire for biblical adherence. Perhaps there was a time when that case could have been made, but with the exception of Jerry Falwell Sr., who died long before the Trump evangelical was born, all of these men have utterly reversed their positions in favor of Donald Trump.

After the *Access Hollywood* tape of Donald Trump leaked in October 2016, Ralph Reed, who was quoted in this chapter saying "character matters" in his condemnation of Bill Clinton, had a far more pragmatic view of the situation.

In an email to the *Washington Post*, Reed referred to the contents of the recording as "disappointing" but ultimately dismissed the idea the recording should impact his endorsement of Trump, saying, "People of faith are voting on issues like who will protect unborn life, defend religious freedom, grow the economy, appoint conservative judges and oppose the Iran nuclear deal." Translation: Character doesn't matter now because voters don't care.

In his endorsement of Trump for president, James Dobson, who I quoted explaining in painstaking terms how the ends cannot justify the means, said, "I believe he is the most capable candidate to lead the United States of America in this complicated hour." Translation: the ends justify the means if the situation is "complicated" enough.

Pat Robertson, who was so disgusted by Bill Clinton that he thought resignation was "too easy an out," also endorsed Trump, saying to him, "You inspire us all." Translation: revulsion can become inspiration depending on your political affiliation.

Hypocrisy in politics, and also in religion, is not hard to come by. It's rampant and poisonous but has become such a mainstay that Americans practically don't notice.

None of these men are unaware of their glowing hypocrisy, and none of them care. And from what I've seen, they've not instilled in their flocks any sense or guidance to prepare them to second-guess these things. Quite the opposite, as I've demonstrated.

But the moral problems associated with this open-eyed hypocrisy are far worse than merely looking to defend behaviors they once decried or their newfound penchant for moral relativism. Donald Trump may be a bellwether for the decline of Christian leadership among evangelical luminaries, but he is not the cause.

As I said earlier, the blatant hypocrisy and selective condemnation from evangelical Christians is a long-standing stigma that has followed the movement for decades. For many, the idea that they would adopt a partisan stance even and in spite of biblical disobedience is no surprise.

But as we've seen, evangelical leaders have gone much further than simply attempting to excuse uncomfortable hypocrisies or rationalize bad behaviors. In the case of Donald Trump, they have gone to great lengths to paint him not merely as an ally but as a vessel of God's will, a submissive and humble servant fulfilling God's purposes.

This devotion to the theology of Trump as God's vessel has the added effect of trickling down into the rest of the party that Trump leads, so long as those party members have demonstrated devotion to this president's agenda.

Consider the support of Roy Moore's run for senator in Alabama in 2018. An outspoken evangelical, Moore was first made famous by defying court orders to remove a monument to the Ten Commandments between 2001 and 2003.[10] Moore came under fire following a series of credible accusations by women that he had engaged in inappropriate behavior with teenagers. One accuser had been under the age of sixteen at the time, which classified the encounter as a sexual assault.

Had those who continued to support Moore in spite of these allegations simply claimed they didn't believe them, the defenses offered wouldn't have been quite as jaw-dropping. As someone who finds the accusations credible, I might merely have thought his supporters were naive or blinded.

What was so galling about the rationalizations for his continued support among evangelicals was that many seemed to accept the possibility that he did everything he was accused of and that, in keeping with the moral relativism the New Good News offers, it didn't matter in light of "greater concerns."

Terry Batton, head of the Christian Renewal and Development Ministries in Georgetown, Georgia, claimed there was no comparison between opposition to Bill Clinton and support of Roy Moore because, "with Bill Clinton, you had immorality in what he stood for, and with Roy Moore, you have a godly man whose positions live out his biblical precepts."[11] In other words, even if Moore did these things, it's merely an example of how he is a godly man who has made mistakes, whereas Clinton's entire persona is defined by his immorality. Basically, Moore is a good guy who may have done a few bad things, but Clinton is just a bad guy.

In Batton's mind, that appeared to be a reasonable approach to saying he would have supported Moore.

David Floyd, pastor of Marvyn Parkway Baptist Church in Opelika, Alabama, did not deny the possibility that Moore committed these crimes. Yet, Moore did not lose Floyd's support. In offering a defense of Moore, Floyd remarked, "All of us have sinned and need a savior," adding, "What I know is Judge Moore believes abortion is wrong. His opponent believes it's right."

Being pro-life myself, I agree that the question of abortion is an important one. But Floyd's calculus is flawed in that he's pretending as though there is a looming moral imperative that simply doesn't exist.

No one was asking Floyd to support Moore's opponent. But modern politics, the evangelical movement, and the GOP at large have designed a rhetorical hierarchy that causes every decision, every moment, every vote, to live under the cover of an ethical dilemma: a choice of "lesser evils." That reasoning is undoubtedly the death of all standards, since everything, no matter how despicable or dishonest or immoral or ungodly, is cast as subordinate to the larger concerns associated with "helping" God achieve His ends.

That divine nationalism that evangelicalism had been articulating all the way back to the era of prohibition and up through the era of the Moral Majority had always been premised on the idea that upholding God's standards was at least part of the mission. The New Good News stated that our nation's divine purpose was so vital that there was essentially no violation of God's moral laws that superseded it.

God has divine purpose. America's role in fulfilling that purpose is uniquely important. The Republican Party is the only party in our divine nation that is aligned with that purpose. The Republican Party can only carry out that purpose if it wins elections.

What's a few glaring immoralities and ruined lives next to stakes that have been set that high?

Defense of Moore was not limited to random lesser known evangelicals. While not an express endorsement, Franklin Graham offered what sounded a lot like the basis of the very moral relativism those lesser known evangelicals were offering. "The hypocrisy of Washington has no bounds," Graham said. "So many denouncing Roy Moore when they are guilty of doing much worse than what he has been accused of supposedly doing."

That's a lot of hedging for one quote. But what is remarkable is how this powerful evangelical voice, who has offered unwavering support through every Trump misstep and scandal, is not only attempting to provide cover for Moore by suggesting that the accusations might not be true, he is concluding that sexual crimes against children are "not as bad" as the hypothetical sins of those in Washington who dared to be offended. Does he think all of Moore's critics are child abusers? If he does, he is living in a bizarre and malicious delusion. If he doesn't, it's a shameful moral equivocation between "child abuse" and some vague nonspecific "hypocrisy."

The "if this is true it isn't that bad or at least not as bad as the actions of others" defense from evangelicals was actually more common than some may be comfortable remembering.

Pastor Earl Wise, of Millbrook, Alabama, told the *Boston Globe* that he would support Moore even if the accusations were true, adding, "There ought to be a statute of limitations on this stuff." As if that wasn't bad enough, he continued, "There are some 14-year-olds, who, the way they look, could pass for 20."

The silver lining in the Roy Moore debacle was that at the very least he earned less support from evangelical voters than would have been expected for a Republican candidate who had been free of scandal.

Though in terms of silver linings, this one will not go far in repairing the image of evangelicals. While it's true that evangelicals were the only voter group that showed slight signs of a dip in voter turnout, dropping from a 47 percent share to a 44 percent share, those evangelicals who did vote still broke for Moore by a margin of 4 to 1.

And as I mentioned in chapter 1, that 80 percent support of white evangelical voters is also the level of support Donald Trump enjoyed with the same group in 2016.[12]

However, that difference appeared to be enough to cause Moore to lose the election.

So far we've seen how evangelical leaders and supporters of this president exhibit symptoms of a disease that not only predates Trump's rise to power but is also growing worse.

How evangelicals came to this point, or were at least revealed to have always been there, is something best described by expanding our view to the conservative movement at large. Evangelical leaders justify these compromises by arguing that we're in the midst of a life-and-death struggle for the culture.

The Culture War

While the term *culture war* has been used for almost a century, its connection to the conservative movement can most credibly be attributed to the 1992 Republican National Convention in Houston, Texas, where Pat Buchanan spoke from the podium in support of President George H. W. Bush's reelection.[13]

In his speech, he referred to the election as a "religious war going on in our country for the soul of America," further clarifying that it "is a cultural war." Then he said that the change a Clinton presidency might offer is "not the kind of change we can abide in a nation we still call God's country."

A decade and a half later, George W. Bush was poised to be the opposite of what Buchanan warned against. Over the course of his presidency, some issues important to evangelicals had seen tremendous progress.

As Clinton's presidency came to a close, evangelicals across the country echoed what Dr. James Dobson, a nationally recognized evangelical leader and president of the powerful group Focus on the Family, at the time had called the need for an "American spiritual awakening." Born-again Christian George W. Bush became the natural progression of that awakening when he ran for and was elected president.

Many of the postmortem accounts of Bush's victory attributed his success, in part, to the Monica Lewinsky affair and an emphasis on character by those portions of voters who had abhorred Clinton's behavior in the scandal.

Upon Bush's victory, evangelicals saw an opportunity to reassert Christ-like expectations of our leaders in the wake of the moral relativism that had plagued the defenses and rationales of Clinton supporters.

Finally, a president they could admire, who was to them "one of us." Someone who, they could expect, answers to a higher moral imperative. Someone who will do away with the notion that subjective perspective has any relevance to absolute right and wrong.

In 2001, in the wake of the September 11 terrorist attacks, Bush issued a warning to other nations, saying, "And we will pursue nations that provide aid or safe haven to terrorism. Every nation in every region now has a decision to make: Either you are with us or you are with the terrorists."

While controversial to some, his clarity and certainty that absolute evil exists was an exciting contrast to the more pragmatic diplomacy of the Clinton years.

By 2004, having instituted funding and support for faith-based initiatives as well as publicly speaking about the importance of prayer and piety, Christians came to view Bush's unapologetic Christianity as a needed breath of fresh air, reflective of a growing desire by Americans to return to the Oval Office a sense of decency that they believed had been missing in the Clinton years.

The memories of Clinton-era moral relativism and ambiguity that Clinton's supporters used to defend against his absence of character were fresh in the minds of many.

An article from writer Steven Waldman in September 2004 titled *The Real Reasons Evangelicals Love Bush*, laid out many reasons for the evangelical contentment experienced under Bush:

> *Finally, there is the war on moral relativism. For many*
> *evangelicals, the root of all Baby Boomer evil is moral relativism,*
> *the sense that there is no absolute good or evil. So when Bush*
> *so clearly and frequently uses those terms, it has resonance well*
> *beyond foreign policy. When he says Al Qaeda is evil, he is,*
> *indirectly, talking to evangelicals about abortion, gay marriage,*
> *divorce, birth control, loud music, thongs, and anything else*
> *they might think resulted from moral relativism. Moral clarity is*
> *essential for fighting not only terror but American cultural rot.*
>
> *There are other, more pedestrian reasons evangelicals love*
> *Bush. Evangelicals tend to be conservative so they like his policies.*

After all, they mostly voted for the very non-evangelical Gerry
Ford over born again Christian Jimmy Carter. (And, to be sure,
there are many evangelicals who dislike Bush altogether.) But the
connection between Bush and a great many evangelicals is deep
and personal—indeed, it's grounded in their reading of how God
transforms men and chooses leaders.

By 2005, a reelected President Bush, having enjoyed a less controversial victory than his initial 2000 win, had spearheaded a complete party takeover, with Republicans controlling both houses of
Congress.

The moral crusade that had found its footing in the closing days
of the Carter administration with the growth of powerful organizations like the Moral Majority had been pleased with the election
of Ronald Reagan in 1980.

But President Bush offered something new and exciting. He was
not merely an ally of the evangelical movement. He was a credible
leader of it. There was a sense that the cultural tide was shifting in
evangelicals' favor. The idea of a Christian nation was at the forefront of popular conservative thinking and discourse. It was a renewed idea, meant very literally.

Among issues important to evangelicals, few rose to the level
of concern over legal abortion. The Supreme Court's decision in
Roe v. Wade in 1974 had been the catalyst of the evangelical movement's rise to power throughout the Carter years.

It was good news then that from the beginning of Bush's presidency in 2001 all the way to its end in 2009, approval of abortion
had dropped 7 points, from 54 percent to 47 percent,[14] and total
abortions had decreased by about 7.5 percent, according to the
Centers for Disease Control.[15]

But the movement had lost ground as well, on issues like gay marriage, which most evangelicals opposed. Thirty-four percent of Americans supported legalizing gay marriage in 2001, with 57 percent opposing. But by 2009, the approval had jumped 5 percent and disapproval had dropped 4 percent.[16] Even support among evangelicals rose slightly during this time (1 percent).

With regard to shifting Americans' views on this issue, evangelical opposition to gay marriage was increasingly a losing battle, and it seemed that the harder evangelicals fought on the issue, the more polarized and stigmatized the movement became.

By the time of President Barack Obama's inauguration in 2009, the issue had become one of America's most divisive, and though most polling showed a majority of Americans were fine with gay marriage, ballot measures to ban it at the state level were often victorious, often with religious organizations spending a fair amount of money on achieving that outcome.

In other words, even as Christians attempted to shift the outcome through government means, the culture continued to slide away from them. Think of the implications of that.

Conservatives and evangelicals were broadly being seen as angry and perpetually outraged. Bitter even. But it was more than being on the wrong side of a cultural defeat that had created the "bitter and self-righteous" evangelicals whom much of America seemed less inclined to listen to anymore. As the Obama era began, calls for unity under a progressive-leaning administration were scoffed at in light of how many conservatives remembered their own treatment during the Bush years.

Not only had conservatives lost at the ballot box in 2008, it had happened after what they saw as years of slander.

Conservatives had spent the previous eight years being insulted and defiled by their opponents. Mocked and marginalized for their beliefs. They saw the emerging world as working against their cherished values, which meant they might as well have been opposing God himself. Since so much time had been spent crafting an alliance between Christian values and the Republican Party, it seemed many believed God was on the side of the GOP. And if God was on their side, then "who could be against them"?

Only liars, frauds, and fools.

The treatment they'd received from political adversaries began to define everything about not only evangelicals but the entire conservative right.

All negative news about Republicans was bias. *All* calls for diversity were attacks on America's identity. *All* protections of other faiths or even of the rights of those who do not subscribe to religious faith were seen as assaults on Christianity.

That picture of the angry conservative was in stark contrast to those early days of the Bush administration.

Compassionate conservatism, now an almost quaint notion, was the term Bush used to describe his personal governing philosophy, and he is the person most associated with it, though he did not coin the phrase.

He described it thus: "I call my philosophy and approach compassionate conservatism. It is compassionate to actively help our fellow citizens in need. It is conservative to insist on responsibility and results. And with this hopeful approach, we will make a real difference in people's lives."

Many purported conservatives today, and particularly among the ranks of Trump supporters, look on that term, and that idea, with disdain. It's often used sarcastically or contemptuously to describe an undesirable and weak philosophy.

That rejection is not new, though.

Conservative columnist and author Ann Coulter, herself an

ardent Trump supporter until recently, in writing a piece *supportive* of the tenets of Bush's philosophy at the time, simultaneously illuminated why there was an immediate *revulsion* among many on the right.

> I, for one, bolted past indifference straight into loathing long ago. A half-century of useless (and more typically pernicious) socialist programs purportedly to help the poor now produces this Pavlovian response to any invocation of "the poor," or "compassion" in a lot of people. Call us cynical, but we've seen this Judas kiss before.
>
> It's not that we hate the poor; it's that every time the government tries to help the poor it ends up removing the marvelous incentives life provides to do things like buy an alarm clock, get a job, keep your knees together before marriage, and generally become a productive, happy member of society.
>
> But consider that even when George W. Bush was talking about "the poor" he insistently said: "Big government is not the answer . . . It is to put conservative values and conservative ideas into the thick of the fight for justice and opportunity."
>
> Bush also said: "So many of us held our first child, and saw a better self reflected in their eyes. And in that family love, many have found the sign and symbol of an even greater love, and have been touched by faith. We have discovered that who we are is more important than what we have. And we know we must renew our values to restore our country." That's not socialism; it's Christian charity.
>
> Out of the Republican Party's seemingly inexhaustible supply of Bushes and Doles, we may finally have located one who talks about compassion without meaning another horrific federal bureaucracy. He better mean it.

Marvin Olasky, editor in chief of the Christian publication *World*, and conservative author of *The Tragedy of American Compassion*, spoke to an audience at the Heritage Foundation in 2001. In his talk, he attempted to rest concerns as to what compassionate conservatism entails. Specifically, he cautioned against rejecting it out of fears that it is simply "big government Republicanism."

The word "compassion" from the 1960s through the early 1990s was a code word used by liberals. Newspaper articles defined a compassionate legislator as one voting for a welfare spending bill. Those opposing such bills were cold-hearted and, by definition, uncaring.

Some of that bias still remains, but more Americans are learning that compassion means with-suffering, "suffering with" a person in distress, developing close, personal ties. More people are understanding that the problem with the welfare state is not its cost but its stinginess in providing help that is patient; help that is kind; help that protects, trusts, and perseveres; help that goes beyond good intentions into gritty, street-level reality.

After making the case that compassion as a matter of government intent need not subscribe to liberal ideologies that favor expensive spending initiatives, Olasky concluded with an appeal to the audience, saying, "It's time right now to say to our fellow conservatives, tear down the wall that sometimes has separated our minds from our hearts! Warm hearts and tough minds, working in unison, can transform America."

Some, like Edward H. Crane of the libertarian Cato Institute, rejected the philosophy in total. In a June 2001 policy report, he described how compassionate conservatism is potentially just a game of outspending your opponent.

It is possible that President Bush is simply naive about the consequences of compassionate conservatism—that he doesn't realize funding will be determined politically and not by merit. Or that even if it were by merit the recipient organizations would be corrupted by a growing dependence on federal funds. Or that those funds will inevitably come with strings attached.

The real danger lies in the casual acceptance of the idea that the federal government should have an "active" role in every-day American life, that if there's a problem, why, the federal government will find some worthy organization to solve it. This is bound to undermine what little principle remains in the Republican Party today. Education is a case in point. After decades in the wilderness, the GOP regained control of Congress in 1994 with a platform that called for abolishing the Department of Education. And why not? There is not a word devoted to education in the Constitution, which means that under the Enumerated Powers Doctrine and the Tenth Amendment (for those too dense to understand the former) education is a responsibility of state and local government or, preferably, of no government at all.

That was then. This is now. Today we are faced with compassionate conservatism. So I was not surprised when I received a fax the other day from the Republican Policy Committee in the U.S. Senate boasting that "since Republicans took control of Congress in 1995, federal education spending has exploded." The headline: "GOP Outspends Democrats on Education."

Ultimately, Bush won the election and, whether it was an embracement of his philosophy or a rejection of the Clinton years, his inaugural address made clear his intent to institute this philosophy into his agenda as president.

America has never been united by blood or birth or soil. We are bound by ideals that move us beyond our backgrounds, lift us above our interests, and teach us what it means to be citizens. Every child must be taught these principles. Every citizen must uphold them. And every immigrant, by embracing these ideals, makes our country more, not less, American.

Today we affirm a new commitment to live out our Nation's promise through civility, courage, compassion, and character. America at its best matches a commitment to principle with a concern for civility. A civil society demands from each of us good will and respect, fair dealing and forgiveness.

The terrorist attacks on September 11, 2001, certainly altered the tone of Bush's presidency, as America quickly moved into a position of defensive strength and unambiguous recognition of evil. But even as the unfolding case for the Iraq War played out, Bush still seemed intent on keeping his promise of compassionate conservative governance.

The results of his efforts are mixed.

Among the chief criticisms from conservatives regarding the deployment of compassionate conservatism: the size of the U.S. government grew by 53 percent over the length of Bush's presidency. This increase certainly can be explained to some extent by the cost of two wars, as well as the 2008 financial crisis and the TARP bank rescue program.

But social entitlement programs were arguably most in line with compassionate conservatism, and they added over $500 billion in entitlement costs as a result of the Medicare Prescription Drug, Improvement, and Modernization Act of 2003.

This same spending criticism, as well as conservative concerns about federal control, was seen in reaction to Bush's education initiative "No Child Left Behind," which many saw as granting bureaucratic control to the federal government through the promise of yet more federal dollars.

According to the National Census Bureau, poverty increased during the Bush years from 11.7 percent in 2001 to 13.2 percent in 2008. However, his promise to help poor countries remained a centerpiece of his efforts to govern with compassionate principles.

Launched in 2003, the President's Emergency Plan for AIDS Relief (PEPFAR) was reported to have saved nearly 12 million lives in its fifteen years of existence.

It is fair to observe that failure of any of these initiatives is not necessarily a condemnation of the underlying philosophy of compassionate conservatism, any more than the successes are proof of its worth.

What is of note, however, is how the perception of Bush has changed in the years since he reentered the private sector. Over the course of his presidency, Bush battled the impression from the right that he was a big government spender hoping to buy goodwill from the left by appearing compassionate, almost as often as he battled the left's portrayal of him as a ghoulish and racist neocon, hiding behind compassionate words while fueling the machinations of war.

But in a trend reversal that surprised some, the left's impression of him has softened in the stark contrast of the Trump presidency (Bush currently enjoys a "favorable" view among the majority of Democrats, climbing from 11 percent approval in 2009 to 54 percent approval in 2016), and overall his approval has climbed to be in the majority within both parties as well as among independents.

However, while Bush's unfavorable ratings plummeted among Democrats between 2015 (the genesis of the Trump-dominated news cycle) and 2018, going from 71 percent unfavorable to 41 percent, Republicans' unfavorable view of Bush tripled over the same period (from 7 percent to 21 percent).

There is a genuine distaste among conservatives for the idea of compassionate conservatism. And not simply a "different kind of compassion" or a "distaste for money being the equivalent of spending."

No, I have come to believe that years of training and the explosion of both hypersensitivity and hyperbolic outrage, combined with unprecedented access to one another through social media (what I like to call "spending too much time together") has fostered a bitterness among those on the right that found many choosing an absurd but amorally rational solution: Donald Trump.

Long before Donald Trump came onto the scene, though, the ingredients required for his ascent had been brought together. It was a toxic mixture of resentment and desire for revenge which spawned the Trump evangelical movement. Evangelicals were tired of being on the outside. It was time to start winning again.

Chapter 4

TIRED OF LOSING

Trump evangelicals are very fond of binary choices. For instance: you're either for Trump or you're prochoice. As a Christian and a conservative who works in politics, I'm regularly exposed to false dilemmas—often presented by Christians—that reveal the depths of intellectual dishonesty being deployed by the right in favor appealing to almost no one new while pushing the persuadable away from agreement. One such false dilemma regards President Trump's border wall and immigration reform plans, which Trumpists use as a Rorschach test to determine people's secret loyalties. You either support the wall or you're in favor of open borders. Essentially, if you detest Trump's immigration rhetoric or, even worse, oppose the construction of a southern border wall, you are clearly an open border fanatic who wants our nation to be destroyed.

National sovereignty is an important aspect of a nation's success, and there is little fault to find with people who defend the importance of preserving sovereignty utterly. For our purposes, and for the purposes of the Trump movement, sovereignty is the broad notion of an impervious barrier around, and total dominion over, a particular domain. Which is to say, sovereignty is the idea of our borders and their relative inviolability, and our resulting independence from the needs and dictates of the rest of the world.

It is a psychologically vital concept to this base of voters. A lot of "liberal" or "progressive" pundits and politicians fail to understand that. It's the expanded self-image of this particularly American voter; *we are separate and independent of you and your needs as well as your assistance, judgment, or opinion.* America the cowboy. America the loner. America, don't mess with it. Sovereignty is the "get off my lawn" of political constructs.

Still as understandable—and in many ways beneficial and accurate—as this point of view is, it is not itself sovereign or inviolable. The concept must be subject to scrutiny and, for Christians, to religious examination.

As a Christian, I believe we must view the issue of our American borders and citizenship, as well as the nature of sovereignty itself, through an appropriate religious lens.

Because we are not afforded the luxury as Christians to say (as Trump did), "Well, I don't bring God into that picture." He must be part of every picture. Putting God on a shelf to wait while man makes decisions "pragmatically" or emotionally is a doomed calculus. Embracing that unbreakable sovereignty must therefore be, to a Christian, dependent upon the answers to some questions, to that religious examination and reflection.

One might ask, for example, Why care at all about this? Why not have open borders and invite everyone in, trusting that God will provide the loaves and fishes to make sure all can eat?

The answer to that question cannot be one of domination or supremacy, nor even jealous care for our own riches.

The definiteness of our independence and our borders must rest on higher principles. It is through our sovereign authority over our shores that we achieve the freedom of our citizens, that we can provide for a common defense, and that we promote the general welfare. This is not a question of purity or protectionism; it's one of national identity as it pertains to a shared set of values.

People often say that America is an idea, not a place. But that idea withers and dies without a place it can reside and defense to protect it.

On those principles, foundational to our republic, it seems clear that a sovereign land, a defined border, an identifiable difference between citizen and noncitizen, are at once utterly desirable and harmonious with principles like freedom of trade, freedom of movement, and a process for immigrating and joining.

Thus, a reformed immigration process that does not cut us off but also retains the integrity of our process is worth protecting from collapse under the weight of internal disorganization or external threats.

The case I am making here is that the impetus need not be exclusion. This is not a book about immigration policy, but it's important to make the distinction between the urge to retain a national corporeal reality and the more base and fearful idea of isolation for the sake of purity.

Important, that is, in the context of Christians on the right.

It is doubtful that Jesus, weighing in on American sovereignty, would advise us to leave the proverbial horse untied. But it seems equally unlikely that He would have called for us to depart from the world, cease to be an example and witness, and hunker down protecting our treasure.

We must balance our reason with our compassion, in other words, because that is the Christian thing to do. You need both, because one without the other is a recipe for disaster. Unfortunately, while even the most decent and well-meaning person might find the balance of reason and compassion challenging to maintain, conservatives and evangelicals have dug so deeply into their new version of reasoning that compassion has become something of a bad word.

This wasn't an overnight change.

In 2013, upon becoming disenchanted by the lack of compassion I saw in regard to Assad's Syrian regime gassing children, I wrote an article calling for what I rebranded as "empathetic conservatism," a sort of outreach, if you will.

If compassionate conservatism was a dud, and it was, it was because the notion was essentially a lie. It offered the governmental philosophy of liberals with conservative overtures about responsibility and morality. It grew government, basically, by appealing to small government people. Or trying to.

Empathetic conservatism, I argued, meant speaking to people in a way that is optimistic and helps them see a brighter future without abandoning our principles or compromising our beliefs. Without that language of empathy, "we are destined to continue losing," I wrote.

Looking back at how the reverse of my prediction came true, it's depressing how much I underestimated the seduction of outrage.

Naturally, many things have changed since 2013, including me, but I still think that a movement incapable of offering empathy will die. Despite any short-term victories that might be attained through fear and outrage, it is critical for the long term that conservatives learn to communicate in a way that is considerate and thoughtful and, most important, reflective of the actual good intentions and outcomes being offered.

In 2013, this idea was on my mind precisely because of seeing the opposite in practice. It was at a conservative event called the *RedState* Gathering.

RedState is a conservative news and opinion website and blog, where I had a byline. (I was unceremoniously fired from *RedState* early in 2018, as a result, I have come to believe, of my criticism of this president, but that's a story for another time.)

RedState formed back in the early 2000s to counter the explosion of liberal blogs cropping up and gaining popularity.

The site has been influential over the years, helping to launch or assist the careers of Senator Ted Cruz, Senator Marco Rubio, Nikki Haley (who most recently served as Trump's U.N. ambassador), and more. Often this was done through the *RedState* Gathering, an annual meeting where grassroots activists in the conservative movement would mix and mingle with candidates and elected officials. It was a big fund-raiser boost for the officials, and a way to get their names spread across the Internet.

The *RedState* Gathering had grown to such prominence that at the 2011 meeting in Charleston, South Carolina, Governor Rick Perry announced from the *RedState* stage that he was running for president.

In 2013, Erick Erickson (currently the editor in chief at the conservative evangelical website The Resurgent) was still at the helm of *RedState*, which was under the operational control of Eagle Publishing.

At this particular gathering, my congressman, Representative Mick Mulvaney of South Carolina, was set to address the group, which was, in the majority, a conservative evangelical crowd.

Mulvaney had won an upset victory in the Tea Party wave of 2010 against fourteen-term incumbent and member of Democratic leadership John Spratt. A successful businessman and fiscal conservative, he'd won the election by focusing primarily on the fiscal condition of the national debt and the seemingly unstoppable trajectory of government spending and growth.

The opening to Mulvaney's speech was littered with the familiar bumper-sticker slogans one expected at the time, such as referring to the need to "save the country."

But there was a caution in his tone as he explained how to

accomplish this objective, a caution that was as unmistakable as it was unexpected.

"I travel the whole country doing events like this," he began. "And people walk up to me and they say 'What are we going to do, number one, to save the party? And number two, what are we going to do to save the country?', and in my mind, of course, those two things go together because our party really is the only hope to save the country. And I get that question a lot, and that's what I'm going to talk about today because I'm going to answer that question. And I'm going to answer that question in a way that's probably different than you've heard before. Probably a little bit more uncomfortable than you've heard before."

He went on to refer to his coming speech as a "disruptive answer to the question" of how we fix the party and the country. He further cautioned the audience that in the process he would likely "offend a lot of people."

I remember being intrigued and concerned to some degree. His presence at the meeting had been at my behest. I'd worked hard to get him exposure as a candidate ahead of his victory, and we had stayed in touch in the years between. I felt certain that he could speak to *RedState*'s audience in a way that appealed to their values, but I was at least mildly afraid I was going to get yelled at for encouraging his inclusion.

As he went on, I began to see not only where he was going but why he had approached it with such caution. After expressing the idea that conservatism can at times be harder to persuasively describe than liberalism since, in his mind, conservatism appeals to the mind and liberalism appeals to the heart, he launched into a story to highlight why, as he puts it, we are "so bad at explaining conservatism."

"I got a cantaloupe delivered to my office yesterday," he said abruptly. He went on to explain that a congressman during debate on immigration in regard to the DREAM (Development, Relief, and Education for Alien Minors) Act, had said, "For every DREAMer who is a valedictorian there are a hundred of them with calves the size of cantaloupes carrying bags of marijuana across the border."

The congressman Mulvaney was referring to was Representative Steve King (R) of Iowa, who made that remark in an interview he gave to the conservative publication *NewsMax* in July of that year. King's full quote being referenced was this: "For everyone who's a valedictorian, there's another 100 out there who weigh 130 pounds—and they've got calves the size of cantaloupes because they're hauling 75 pounds of marijuana across the desert." King went on, "Those people would be legalized with the same act."

But at the time I wasn't yet sure of the relevance of this story or how Mulvaney intended to use it as a way to identify how to "save America." The reaction from the audience was silence. Mulvaney pressed on, reiterating, "I had people come to Washington, D.C., with hundreds of cantaloupes, get them through security, find my office, and put it on my desk with a note.

"What do you think it took us to push people to do that?" he asked the audience.

"One thing I've learned is that if you push people too far, they eventually will push back," he said. "And I don't think that comments such as 'calves the size of cantaloupes' help us. Did it reach out to independent voters? Did it help us? Did it help us spread the message? Did it make us look like the party that people want to be members of?"

As I surveyed the audience, they looked confused. Perhaps sensing this, Mulvaney summarized his point, saying, "You don't persuade people by insulting them."

To the audience assembled, this must have been a radical prop-osition. Mulvaney himself must've known it would be, given the careful way he'd opened his speech. And indeed, from this point forward, the audience did not laugh as hard at his jokes, clap as vigorously at his platitudes, or seem as connected to his message.

It became especially intense in the room as he made clear that his disdain for the divisive way Republicans approached political discussions was not reserved only for the elected officials talking about cantaloupes, but the grass roots as well.

He regaled the now suspicious audience with another story, this time of an activist guilty of the same rhetorically bad behavior he had identified from Congressman King. This story was about a Tea Party rally that had happened a few weeks prior. He explained that someone near where he was standing was speaking Spanish and that a "Tea Party guy" standing next to Mulvaney yelled, "Hey, you idiot, learn English!"

It turned out the Spanish-speaking gentleman this "Tea Party guy" was shouting at was Republican congressman Mario Díaz-Balart of Florida, who Mulvaney went on to explain was the son of a Cuban immigrant who had fought Castro's regime; at that moment, Díaz-Balart had been speaking to a group of right-wing conservative Hispanic preachers.

I was certain as I watched Mulvaney that he believed he could reach this audience and get them to see what he saw. To get them to understand that compassion is not anathema to conservative ideals.

For many conservatives, who had felt strangled by niceties and the seemingly one-sided expectation that they should be decent toward those they deigned indecent, there already had been a growing disdain for the mere suggestion that compassion should play any role in politics and governance. It was clear to me that they would view calls for compassion—from this stage at this moment from this congressman—as an affront to their widely held belief that "sober truth" delivered as cruelly as possible was the only way to deal with difficult realities.

Following these remarks, the normally raucous crowd found few reasons to offer Mulvaney applause. He received the most when he talked about what a great messenger of conservatism Representative Trey Gowdy had become, and once again, tepidly, at referencing a bill that he voted against. But even-low hanging fruit, such as recognizing members of the audience that he had met prior to speaking, barely registered a reaction from the assembled.

His complimenting of the audience about how they represented the best of what conservatives had to offer received no reaction at all; at that point he shifted somewhat to try to win them, imploring them to understand that he "did not want to give them the wrong impression" and that there was a time and place for "red meat speeches" and "cheerleading," even acknowledging that perhaps he had been "out of line" and that was the type of speech he should've given.

But he beseeched them to consider that conservatives needed people capable of explaining "difficult problems, complex issues in a way that's understandable and then articulate a resolution to them."

As the polite applause at the end of his speech subsided, he took three questions from attendees, with nothing out of the ordinary coming up.

But I remained concerned that I'd screwed up by insisting on Mulvaney's inclusion, despite the fact that I wholeheartedly agreed with his comments. Later that day I was tasked with walking the hallways with a hotel video crew and asking attendees their thoughts on the conference; the answers related to Mulvaney confirmed my suspicions.

To those I spoke with, he was an unwelcome addition who had scolded them about something as trivial as language at a time when we could "not afford" to prioritize such nonsense.

Mulvaney had rightly identified the unnecessary vitriol that had become commonplace among conservatives as an effective way to turn people *away* from our message. And upon relaying that to a largely evangelical audience, he had been dismissed as weak.

To the assembled, there were those greater considerations I mentioned earlier. The need to appear, much less actually be, compassionate was simply a denial of pragmatic reasoning. It was defiance of a desperately needed embrace of objective facts and harsh realities.

But it was more than that, too.

There was indifference, certainly, but there was also a sense of being wronged. They had the idea that conservatives and Christians had been given the short end of the stick for too long. They were "mad, and they weren't going to take it anymore."

This is a very important part of the puzzle.

The right had adopted a persecution complex. They felt beaten down, demoralized, silenced, and maligned for that very embrace of logic and for meeting hard reality with hard facts.

They have an American impulse for the independent and unfettered self, coupled with the pragmatist's grasp of the harsh realities, all packaged together with seemingly clean logic and pure and humane motive. This was the absolute right-wing self-image.

To be seen as anything else was injustice. So each day was injustice, because that self-image was under constant assault by "the left" and popular culture.

To conservatives, all of this was at the behest of or instigated and exacerbated by the mainstream media, Hollywood, and the elites. At the hands of these adversaries, they'd been wrongly and unjustly accused of motivations and animosities they did not believe they held.

All of this was, in one way or another, the result of a culture that was not simply in decline, but that was actively rejecting the essence of being American.

Among conservatives there was no greater feeling of a lack of progress in that culture war than the fact that the decline continued even in the wake of the electoral victories of 2010.

Persecuted

In 2013, Dr. Ben Carson, a highly regarded neurosurgeon and self-professed evangelical, made headlines and launched a political career following a speech he gave on the dangers of political correctness at a prayer breakfast attended by then president Barack Obama.

In it, Carson said, "We've reached a point where people are actually afraid to talk about what they want to say, because somebody might be offended." He insisted that we must "get over this sensitivity as it keeps people from saying what they really believe."

He referred to political correctness as "dangerous" and contended that "one of the founding principles was freedom of thought and freedom of expression. PC puts a muzzle on people."

He further claimed that we as a nation have "imposed upon people restrictions on what they can say, on what they can think," adding that the media is the primary culprit in this as they "crucify people who say things really quite innocently."

These words, among others, catapulted Carson to national fame, eventually leading to his moderately successful campaign to win the Republican nomination in 2015, which eventually resulted in a position in President Donald Trump's cabinet as the U.S. Secretary of Housing and Urban Development. It seemed speaking boldly against political correctness, among other aspects of his speech, had proven to be a powerful message to conservatives who were hungry to hear their views spoken directly to the president.

Mark Hannah, left-leaning author of *The Best "Worst President": What the Right Gets Wrong About Barack Obama*, penned an op-ed for *Time* in 2016 offering enthusiastic support for what he contends are tenets of political correctness. Like Carson, he cites founding principles to support his view, saying, "Political correctness is a longstanding American tradition and a deeply rooted value." He added, "Trump can say what he will about Muslims and Mexicans, but thoughtful journalists and pundits can and should say what they will about Trump."

The contrast is explained in part by differing definitions of political correctness. The user-curated *Urban Dictionary* defines it in a way that I'd argue is quite close to the conservative view, calling it "a way that we speak in America so we don't offend whining pussies."

Conversely, I asked a liberal friend for her definition, and she defined political correctness as "the term we assign with a sneer to any request that we grow up and act like decent human beings."

One side sees political correctness as an onerous system curtailing freedom of speech and expression, the other as basic rules governing politeness.

The anti-PC crowd has a slight majority in the United States. A Pew Research Center poll from 2016 found that 59 percent of Americans believe "too many people are easily offended these days over the language others use."

While that poll may give the false impression that dislike of political correctness is somewhat down the middle, further specificity tells a different story. According to the poll, "eight-in-ten (78 percent) Republicans say too many people are easily offended, while just 21 percent say people should be more careful to avoid offending others."

Democrats, on the other hand, were more accepting of political correctness, with 61 percent saying they "think people should be more careful not to offend others" with only 37 percent saying people shouldn't be offended so easily.

Even a cursory observance of the ways in which Democrats and Republicans approach opposition demonstrates the findings of these polls. For instance, Democrats often claimed that racism was at the heart of opposition to Barack Obama. Conversely, Republicans are more often seen mocking opposition to Donald Trump as "liberal tears." Opinions may vary on the sincerity or the accuracy of the claim, but what is clear is that these days Democrats tend to skew toward indignation while Republicans tend to skew toward ridicule.

Democrats focus on dislike of those who offend, Republicans on dislike of those offended.

The right-wing tendency to ridicule is a tell of how disgusted conservatives are with political correctness, often due to personal experience.

Let me relate this abstract idea to a real-world example: my own.

Like many on the right, I too indulged in outrage as a way of contending with my frustration surrounding perceived injustice. Right-wingers, conservatives, Republicans, Christians . . . we felt unfairly maligned by the left, branded as hateful or motivated by evil. It is not a good feeling. It's frustrating and you feel powerless to combat it.

Outrage was our retaliation, and I was filled with it.

In 2009, after the election of Barack Obama, I found myself enraged by years of combating false accusations regarding my motivations leveled at me by the left. This is not a mere feeling. The left truly did unjustly characterize me, and others, and my anger was a pure righteous rage.

From my point of view, every action President Bush took was met with hysterical mania by the left, which painted him as everything from a racist to a Nazi. I saw marches with signs that read "Bush Is Hitler," and my support for him was often characterized as an embrace of racism or white supremacy. It was infuriating.

Disgusted by these accusations, and carrying around resentment, I reached peak outrage upon seeing a video put together by Demi Moore and Ashton Kutcher in the wake of Obama's election.

This cloying video, which to this day causes my neck to twitch, featured a bevy of famous celebrities taking what they referred to as a "pledge" to support Obama and try to find common ground with other Americans. It was preposterously solemn, condescending, and sanctimonious.

I despised it.

So, on Obama's Inauguration Day, I resolved to no longer "take it" and expressed my disgust in an article for *RedState* titled "My Conservative Pledge." I cannot stress enough how closely my path and that of the future "deplorables" were aligned at this time. You can hear it in every word below. I can think of nothing more instructive than my own example having come out the other side on . . . the other side.

To the pledgers in this video:
   Where were you for the last eight years? Where were you when
we were attacked on 9/11? Where were you when the world agreed

that Saddam had WMDs? Where were you when the media was lying about our president? Where were you when the Iraqis were suffering? Where was your compassion for them? Where was your desire to be amicable with people you disagreed with these last eight years? Where was your "understanding"? Where was your friendship? Where was your bipartisanship? Where was your hope? Where was your restraint?

For eight years I've seen hatred, loathing, anger, ignorance, violence, suppression, oppression, blacklisting, and all together a bunch of meanies. I've seen Bush called a Nazi. I've BEEN called a Nazi. I've been told that I support blood for oil, that my beliefs are archaic, and that I'm a racist. I've seen my God spit upon. My beliefs mocked. My heroes destroyed and my integrity impugned. I've seen my country's soldiers used and spit out for political gain. I've seen marches and been told that dissent is the highest form of patriotism. I've been told that everything I stand for and believe in is murder, greed, and fascism.

And now. After all this YOU are pledging to me that you will change because of one man? One man will make you different and cause you to be reasonable and you fully expect that I should and will reciprocate? You expect us all to come together after this and act like the last eight years didn't happen? As Obama steps forward and has the same reactions to Gitmo & Iraq and NOW you understand the complexities? Now you can have intelligent debate without insulting everything I stand for?

I will make a pledge to you. I pledge to spend every free moment working to counter your efforts. Not simply for the sake of countering them, but because they are not what our country stands for. They are not less government. They are not liberty. They are not justice. They are not FREE. I will not, I cannot allow compromise on these issues. There IS no compromise for me on abortion. There IS no compromise for me on terrorism. There IS no compromise for me on socialism. There IS no compromise for me on much that you stand for.

Unlike you I will not fight unfairly. I will not lie about you. I will not toss insults carelessly and endanger my country and its brave soldiers. I will not hope for a bad economy. I will not invest myself in my own defeat. I will not call our president a Nazi, or a Stalinist. I will be a person of reason. But I will be relentless and unforgiving. I will use my mind and my faculties and everything at my disposal to squash this ideology that your anointed one has brought to bear. I will work to protect this country's ideals and prevent the destruction of our morals.

You created me these last eight years. I pledge that I will become your worst nightmare because of it.

I ended it with my full name, which marked the end of any anonymity I had previously enjoyed when writing online. The anger I felt about the accusations and disrespect I'd endured was public, along with my name. If I had to identify a word to best describe what I sought, that word would be *justice*.

You see, to me and—as it is the purpose of this anecdote to illuminate—many others, it had been an ongoing and intolerable injustice that our political views and policy ideas were viewed as a permission slip to impugn our very character as human beings. The left seemed incapable of good faith, and increasingly we on the right identified "political correctness" as the weapon they used to bludgeon us into silence.

I relate this now not to air a grievance but to demonstrate the state of mind that so many people, myself among them, were in at the time of the transition from Bush to Obama.

As Obama's presidency continued, this state of mind was continually confirmed. Time after time, we saw that the left would not deal in good faith but would still demand it of us.

Joining Twitter that year only helped clarify and harden this point of view for me. For my own experience, vile leftist trolls regularly came after me for columns I wrote. Rarely was this in the form of meritorious or fact-based debate. Rather it was typically and routinely the simple assumption that I must be racist, or hate the poor, or entertain some other atrocious, immoral perspective.

This was widespread. It was especially true in the first year of Obama's term. To dispute him was to hate him, and to hate him was to hate all but white men. This was the operating premise, and your reasons for your beliefs were absolutely immaterial.

In other words, as far as the activist right was concerned, this was political correctness run amok. The left had made the rules, and the Democrats, the mainstream media, and Hollywood all worked together to enforce them. Not through a fair application of those rules, but through a political litmus that cynically used victimization to deflect the possibility of reasonable debate.

It was one-sided and unforgiving. One article demonstrates the way the left is quite aware that political correctness is not some invention or imagined grievance from the right. It's real, it's dangerous, and when it is wielded against the left, they take notice and demand people knock it off.

Liberal columnist Jonathan Chait's "Not a Very P.C. Thing to Say," written for *New York* in 2015, elaborated on what many conservatives would say had been their own view of how political correctness was a weapon against intellectual discussion.

> Under p.c. culture, the same idea can be expressed identically
> by two people but received differently depending on the race
> and sex of the individuals doing the expressing. This has led to
> elaborate norms and terminology within certain communities on
> the left. For instance, "mansplaining," a concept popularized in

2008 by Rebecca Solnit, who described the tendency of men to patronizingly hold forth to women on subjects the woman knows better—in Solnit's case, the man in question mansplained her own book to her. The fast popularization of the term speaks to how exasperating the phenomenon can be, and mansplaining has, at times, proved useful in identifying discrimination embedded in everyday rudeness. But it has now grown into an all-purpose term of abuse that can be used to discredit any argument by any man. (MSNBC host Melissa Harris-Perry once disdainfully called White House press secretary Jay Carney's defense of the relative pay of men and women in the administration "mansplaining," even though the question he responded to was posed by a male.) Mansplaining has since given rise to "whitesplaining" and "straightsplaining." The phrase "solidarity is for white women," used in a popular hashtag, broadly signifies any criticism of white feminists by nonwhite ones.

Chait's take was viewed by conservatives mostly as a "well, duh" moment. This is what we'd been screaming about for years. What seemed clear was that the left was only catching up as a result of tumbling down the slippery slope they'd eagerly lubricated.

National Review columnist Kevin D. Williamson said of Chait's newfound disdain for political correctness, "Jonathan Chait's recent critique of political correctness insists that the phenomenon has undergone a resurgence. It hasn't; contrary to Chait's characterization, it never went away. The difference is that it is now being used as a cudgel against white liberals such as Jonathan Chait, who had previously enjoyed a measure of immunity."

At the *Federalist*, Robert Tracinski remarked of Chait's column that "once a system is in place and its basic principles are established, it tends to keep operating to the logical end point of those principles. And the logical end point is exactly what Chait is whining about: Binary Persons Without Color on the left now face being summarily labeled and dismissed as bigots—the very same treatment they have so eagerly applied to the right for so many years."

The impatience with Chait is understandable. He was highlighting a problem that conservatives had raised as though there was no precedent and asking for a return of good faith that he showed no indication he was willing to offer others. As Williamson noted, it seemed more like he was upset that the weapon he was happy to use against others was suddenly directed at one of the "good guys."

Jonah Goldberg, writing for *National Review*, shared the feeling of schadenfreude but also offered a measure of caution in dismissing Chait's criticism of political correctness: "Maybe he can withstand being unfairly called a racist and sexist because he refuses to buy into the agenda of those who would use such accusations to silence dissenters from their agenda. If he does, I say again, good for him. And, if in seeing the intellectual bankruptcy in such accusations when hurled at him he realizes they are very often equally bankrupt when hurled at his opponents to the right, that would be even better."

Of course, that realization never came, and the left continued in their debasement. This failure to recognize or criticize their own actions and arrive at a new standard had consequences that reached far into the future, all the way to the 2016 election. The image of a pendulum swing in politics and culture is well established, but in this case the more appropriate visual is that of an equal and opposite reaction. The years of being called racist or other monstrous things took a toll, a frankly understandable toll. Unfortunately, the reaction was itself a problem and, as "equal and opposite" dictates, a rather big one.

As Mick Mulvaney predicted at the *RedState* Gathering, conservatives and evangelicals had become overly tired of being characterized as evil. Had become so exhausted, in fact, that many simply became that which they'd been named.

The pushback, then, wasn't an intellectual rejection for many. Instead, it was a petulant embrace. A willful surrender, so to speak, to the evils laid upon them. They became the caricature.

So it was that by 2015, accusations of racism or sexism on the right did not even merit consideration, by and large. They were rejected out of hand. They were, without examination, to be seen as mere weapons of identity politics. Talk of victims or discrimination would often result in the rolling of eyes at the least, or just as often angry fighting back. Always, dismissal of particular circumstance was a foregone conclusion.

Yet it went even further than that. By late 2015, any expressed concern at all for fellow human beings was labeled "virtue signaling," a new cardinal sin on the right. Any desire that we offer decency even in the face of indecency was ceding the battleground. Turning the other cheek had become surrender. Instead, we were to answer the low road *with* the low road, ostensibly until our enemies "got it."

If there's one common theme among the criticisms currently being leveled at the right, it's that they are devoid of compassion. In fact, there's often no better way to get conservatives to walk away from a discussion than to ask them, "But is what you're suggesting kind?" Such a question would be viewed as offensive, if not treasonous, to the cause. One bristled at the notion that "kindness" should be a societal consideration, because it somehow implied that you were not inherently kind. There was simply no space in which human compassion and hard reality could coexist any longer.

The politically correct culture had seemingly robbed conservatives of the will to be "better." Appeals to being better, or above, or moral, or just—any of these were the virtue signal sin.

During Trump's candidacy, many conservatives openly mocked those who called for decency or an adherence to principles, using the Internet's withering tools of contempt. Memes were made, nicknames repeated. The war was on, and now it wasn't just against the left, it wasn't just about accusations or injustice, it was directed at the right as well, against anyone who suggested anything less than total warfare.

"But muh principles" became a mocking refrain, shorthand to dismiss an argument in total, perhaps accompanied by an edited photo of conservatives being marched into death camps.

I wish I were exaggerating, but I'm not. Much as the left had spent decades using the word *racist* (sometimes rightly applied, sometimes not) to label and shut down opponents, so the new right used *virtue signal* or *muh principles* or other epithets and sayings to shut down the moderate right by the angrier right. The more right-wing right.

This was early purge behavior. Later, and on into Trump's presidency, the purge behaviors became far more overt and direct.

The worldviews prompting this persecution complex and its attendant backlash were held together by a hodgepodge of reasoning consisting mainly of examples of outrages committed against the right, and along with these outrages came an ever-increasing insatiable desire for things to be outraged by.

Many conservative websites, such as the Gateway Pundit and Breitbart News, were happy to peddle this outrage, seeking clicks and traffic by seizing on the modern conservative's sense of injustice. As the clicks came, and the ideas were reinforced through group dynamics, they became even more pronounced. Anger had become a currency.

Gone was compassionate conservatism, and here, poised to replace it: the new era of anger.

Anger

This was now the default position among conservatives, whose faiths may vary on the edges but are Christians 85 percent of the time.

For a variety of flavors of conservative voter, and especially those thought of as evangelicals, the perpetual sense of outrage, combined with an overwhelming sense of persecution, manifested as an easily recognizable self-righteous bitterness.

Perhaps it is understandable or even inevitable for members of any identity group to become embittered if they perceive that the world around them rejects both the validity and value of their ideas. In the case of Christian evangelicals, this combination of self-righteousness and bitterness makes for an especially unappealing sale to anyone not already part of the movement (and a fair amount within).

You might think evangelicals would remember that the Bible warned, "Everyone who wants to live a godly life in Christ Jesus will be persecuted." Despite often quoting this verse, the evangelical right seems almost dumbfounded that things have not gone their way. Not just dumbfounded but cosmically offended. Far from believing it is the burden of their faith, they are outraged that they must endure this hardship. How different this attitude is from Paul's instruction in James to "count it all joy, my brothers, when you meet trials of various kinds, for you know that the testing of your faith produces steadfastness."

Wallowing in acrimony and wounded pride, many seem to

forget that the Bible also says to "delight in weaknesses, in insults, in hardships, in persecutions, in difficulties." Instead, in the modern evangelical movement the active and vocal members seem to embrace bile and uncontainable outrage over America's apparent cultural decline. Not to mention in no small part resentment over the decline of their own personal status in the world.

Feeling defeated, downtrodden, or cast aside is not an easy thing to cope with, but then that's what study and prayer and fellowship are for, right? To guide our path and remind us to lean into our faith?

But all too often we see something darker.

It was there in the reactions to Representative Mulvaney's appeal for higher discourse.

It was there again and again and again in the debates over immigration, especially as it related to children of illegal aliens.

In fact, few things show the bitter conservative worldview in more perfect relief than the debate of immigration policy.

Looking backward again, you will recall that in 2014 there was a growing problem at the border.

Sixty thousand youths had crossed the border illegally, which was overwhelming border patrol agents and pitting border governors against President Obama in a fruitless but acrimonious blame game.

Bob Dane, executive director of the radical Federation for American Immigration Reform (FAIR), captured the essence of the right-wing view, saying, "The surge, and the collapse of immigration enforcement nationwide, has been brought about by the Obama Administration's six-year scaling back of enforcement—changes that have incentivized millions thinking of coming to America illegally."

He went on to say that "illegal aliens, anxious to capitalize on the administration's offer of a free pass into the U.S. place their lives and their children's lives" in danger. His one concession to compassion is this: "Some, unfortunately, die along the way."

An alternative view, one that cites an organic crisis that had been building for years, was offered at the *Huffington Post*:

> *Rising street violence in Central America, along with political instability in Honduras gave people a reason to leave. Previous migratory patterns established since the late 1970s—first for political reasons, then for economic ones—meant many Central American youths have family here in the United States. And a hole in our immigration enforcement system was exploited both by families desperate to bring their children here and by human traffickers eager to profit from their predicament.*
>
> *In addition to being based on a fallacy, the GOP's call to fight this crisis with more border security would do nothing to solve the problem. These children aren't a security threat. They're turning themselves in to U.S. authorities at the border.*
>
> *Some 550,000 people benefit from DACA today. Because of the policy, they can work in the only country they know as home and can live with a modicum of security, knowing that—at least for now—they don't have to face the perpetual fear of deportation. Conservatives want to take that away, even though there's no compelling reason to believe it would stem the border crisis or have any other positive effect.*

In terms of the standard tension that I've already gone to great lengths to describe in this book, the DACA crisis of 2014 is a perfect example.

The right, certain of the objective correctness of a point of view that they perceived had been arrived at rationally and dispassionately, maintained a pose of detached matter-of-factness. "Some, unfortunately, die along the way" was all writer Bob Dane could muster in terms of the difficult situation these migrants faced.

He does not see the status of our borders or our willingness to open the door as contributing to the plight of those seeking our shore. In his view, the danger they face is precisely why they made the journey in the first place. That is to say, of course it's tough out there, that's why they want in here.

On the left, the *Huffington Post*, while not denying that the journey was in an effort to get to America and the "open arms" being offered, focuses on the feelings. We should be considerate of the hellscape these people are leaving and make our resources available to assist them.

Both points of view use good principles to reach their conclusion.

Christians must, as we saw on the issue of sovereignty, consider whether their view is a balance of compassion and reason. Christians should make sure they are charitable but not careless, adhering to the philosophy of Christ and to the laws of Caesar.

That would be the moral path.

For my part, I struggle to imagine a scenario where the most affluent nation in the world, which purports to be a Christian nation, turns away those who suffer or seek aid. But leaving my own view aside, it is unquestionable that Christians must consider the morality of the issue and not just practicality or, even more bleakly, the expense.

So there we were, this Christian nation, pondering morality and practicality as this crisis of young asylum seekers and would-be immigrants unfolded at the border. In that moment, the most fascinating and poignant role was played by radio host and Fox News personality Glenn Beck.

Beck traveled to the border personally at the time, not to take a stand against illegal aliens being granted entry but neither to demand they be allowed to stay.

He traveled to the border for charity's sake, and once there he began giving out care packages and toys to the children as a way to at least make what must be a frightening ordeal just a little less overwhelming.

At the time, he said regarding the children, "Through no fault of their own, they are caught in political crossfire," adding, "While we continue to put pressure on Washington and change its course of lawlessness, we must also help. It is not either, or. It is both. We have to be active in the political game, and we must open our hearts."

The response from Beck's audience and from conservatives across the Web revealed the gross extent of the self-righteous, amoral "rationalism" in which the right was indulging.

Beck told audiences and the press he'd begun to receive violent emails and threats, with some claiming he'd "betrayed the Republic" with his charitable act.

Gossip website Gawker collected some of the vitriol and hateful commentary Beck faced for his actions and published them.

Even less defensible were reactions among Breitbart News commenters, who couldn't believe Beck would use even private funds to make a young undocumented immigrant's life easier. The following are comments published in response to the story of Glenn Beck handing out toys to children:

- What a disgrace is right! Send every illegal back & stop the freebies! And once again Obama wants the taxpayers to pay for his screw-up!
- If bums started hanging out in your yard would you feed them?
- Beck should be using his "pulpit" to argue strongly for immediate deportation and for focusing on the LAWLESSNESS of the

Administration. One final point is . . . just like with the Gov't Freebies given to these illegals . . . and the Tacit Amnesty that Obama is promising with a Dream-Act like Executive Order . . . Beck's stunt just serves to ENCOURAGE MORE illegals to come . . . which is ultimately Destructive to these many children.

Republican state senator Andrew Brenner of Ohio (at the time a member of the state's house of representatives) also objected, calling it amnesty and saying of Beck on Twitter that he was "helping the illegals out."

"How is that different than giving their parents jobs here?" he asked. He then asserted that helping children who had crossed the border illegally was itself an illegal act. This from a legislator.

When questioned on how Beck was breaking the law, Brenner doubled down, saying, "So anyone who feeds, gives shelter, water and food to illegals isn't breaking the law?" At no point did he consider that the source of his rage was that children were getting toys from a charitable organization, not the government, while they waited to find if they were going to be shipped back to the place they'd risked dying to escape.

This is not a moral point of view. It is vicious. It is not a rational point of view. It is unreasoning.

The same was true of the right online in general. Individual activists and regular people on social media were equally outraged. Twitter user TaylorinNY, whose bio read "Christian," said of Beck's decision, "Disagree, Beck, there are Americans who need our help 1st."

Also on Twitter, user Libertytombob remarked, "Assisting Illegals, other than basic life support, is like negotiating with Terrorists, U just encourage it."

Breitbart News writer John Nolte employed a strategy common

on the right in defense of what appears to be uncompassionate behavior.

> *It's just a fact that these children are already being cared for once they arrive here. I'm not even sure they need Beck's charity. Regardless, the unthinkable danger for them occurs during the journey across a continent to get here. It's not just the natural elements these children have to worry about, I'm hearing on the news that fully one-third of young girls are sexually assaulted during the trip.*
>
> *Therefore, the truly compassionate thing to do is to ensure you're not doing anything that might encourage more parents to send their unaccompanied children on that harrowing trek. And it is not insane, unreasonable, or lacking in compassion to argue that news of a major American media figure greeting children at the border with toys could be used by the drug smugglers and human traffickers already exploiting these kids as a way to recruit more.*

The idea here is that Nolte is being cruel to be kind. It may appear I'm being mean, the reasoning goes, but actually I'm working toward something larger, which is intended to help you. Or to put it another way, choosing an apparent evil in service of a greater good. Sound familiar?

An analogy would be spanking your child for trying to cross a busy road. Certainly it may seem cruel to the child in that moment, but it's in the service of something larger: that child's safety. This analogy falls apart, though, when you consider that no amount of kindness from the parent means they won't spank them. Kindness and punishment go together, in fact, because both impulses are motivated by compassion.

As Dana Loesch has said, "Mercy and justice can coexist." You can uphold the law and rewrite it in such a way that may prevent the types of problems seen in 2014, but there is no requirement that you withhold kindness throughout.

This theme of heartlessness as the only way to combat lawlessness permeated the entire movement, and by 2015, the self-righteous bitterness that ruled the right took us into a campaign season prepared for only one outcome, though many of us who should have known couldn't see it coming.

Chapter 5

THE ALTAR
OF WINNING

Machiavelli is the only political thinker whose name has come
into common use for designating a kind of politics, which
exists and will continue to exist independently of his influence,
a politics guided exclusively by considerations of expediency,
which uses all means, fair or foul, iron or poison, for achieving
its ends—its end being the aggrandizement of one's country or
fatherland—but also using the fatherland in the service of the
self-aggrandizement of the politician or statesman or one's party.
—LEO STRAUSS

Scholars, experts, and know-it-alls with Google will often say that Niccolò Machiavelli was a man of greater
nuance than the popular perception suggests. This, of
course, is necessarily true of any people—we all exist in many
more degrees and shades than others may witness or remember.

It's not just experts of the caliber of philosopher and historian
Leo Strauss, quoted above, who remind us of that subtlety in men.
Even a high school history teacher in Lynchburg, Virginia, often
told his students that *The Prince* was instructive in the negative.
A bitter satire from which one must learn to "do not" rather than
to "do."

But through time, whether its namesake fully subscribed or not, Machiavellianism has become the symbol of self-interested political philosophy for the minds of the modern West, from the ivory tower to the lowly tweet. "The ends justify the means" is an utterly commonplace philosophical and political notion.

It is usually cautionary, not embraced—especially on the right, and most especially on the evangelical conservative right. Or at least, that was the case before the New Good News.

Perhaps, as with Machiavelli, this cannot be held as the direct intent of Donald Trump, but the intent of Donald Trump is not the subject we are examining here. That Trumpian, née Machiavellian, ideal is very much a deliberate, intentional philosophy on the Trump right, and openly so, and that is to what we look now.

Part of the reason, part of where that begins, is in a rejection of the politically correct persecution of the right by the left, be that merely perceived or real. All of which is to say that for the Trump right, the end of overthrowing the popular left's regime is worth achieving by any means necessary.

For the side so utterly fond of "facts don't care about your feelings" political takes, this justification is motivated by a rather base *emotional* motivation: the heated desire for vengeance.

If Machiavelli was the embodiment of cold calculation, Trump is the personification of retribution.

Having spent years crafting a brand of brash, unapologetic, politically incorrect fearlessness, Donald Trump was the ultimate dream come true for those conservatives who were fed up with feeling scolded and who bristled when decency or politeness were waved in their face. He was the lifeline idea for conservatives who felt they'd been taking it but had never been allowed to dish it out.

In him, in his movement, they saw the chance to fight back. It's a theme they used repeatedly and loudly. "He fights back." No more taking beatings, or being called evil or immoral or racist or misogynist or ... backward. Now for wrath.

The Internet, again, connected the people who felt this way with the people who would use those feelings. Right-wing website Breitbart News was in the vanguard of this coalition of the angry and the aggregators of anger.

Breitbart News was founded by new media titan Andrew Breitbart. Andrew, who was best known for his near-singular focus on disrupting media bias, which he referred to as the Democrat Media Complex, passed away in early 2012, and in the years that followed, the site underwent radical changes in focus. The website had always expressed conservative views and was critical of Democrats and weak Republicans, but in the aftermath of Andrew's death, it seemed clear that the website began to shift from critiquing the mainstream media's agenda to forging a narrative of its own (some would argue, a propaganda mission).

In the period after Andrew's death but before the emergence of Trump, Breitbart News served two primary functions: it catalogued every outrage or slight of the right, and it worked hard to foster a sort of desperate, searching frustration. The millions of readers and the networks those readers influenced were looking for their Barack Obama. The anti-Obama.

But for the manipulator at its helm, executive chairman Steve Bannon, this was not an aimless search. He already had a name: Donald Trump. And slowly he began to direct the editorial narratives and feed news about Trump to the base, perhaps giving the impression to readers that his rising influence was organic.

Breitbart worked hard and deliberately to foster the belief among conservatives that only a Trump presidency would make the vengeance of the aggrieved a reality. Shortly after Trump announced his candidacy, the site ran "The 10 Important Reasons

Trump Would Make a Great President." It knew exactly what it was doing and precisely which buttons to push.[1]

That list is, in its entirety, very telling. But the lantern hung most obviously was the fourth item on the list; number four was the ultimate aphrodisiac of the moment: "Trump is not politically correct; he's not afraid to say what he believes and has ignited an honest debate."

Again, it cannot be overstated the extent to which that message sang in the hearts of right-wing activists and voters.

While Breitbart's support stood mostly alone among conservative publications in those very early days of Trump's candidacy, he essentially led the pack of primary candidates among *voters* from the moment he threw his hat in the ring. Eventually, the publications that rejected him would be picked off one by one with the same efficiency he picked off his primary opponents, leaving only a scattered few openly critical conservative publications today.

Trump beat the anti–politically correct drum, and it was a rhythm conservatives had yearned to enjoy after years of feeling unfairly maligned for just saying what they thought was true.

This dynamic, combined with eight years of, at times, substantial conservative gains but mostly historic losses in terms of legislation and regulation under President Obama, made conservatives ripe for the picking by Donald Trump.

Not picking. No, the *absorption* of the right by Trump. The previous identity was subsumed by the new one. This is not just true in the abstract, it was also true in a real-world sense. People on the right enamored of the Trump movement literally adopted his mannerisms, his speech patterns, his very words.

This donning (forgive me) of Trump's persona has spread throughout the GOP and also caused an alignment with other previously fringe elements such as the alt-right. And of course, you cannot discuss the rise and success of Trump without examining that group.

Though the term *alt-right*, allegedly coined by white supremacist Richard Spencer as far back as 2010, has enjoyed many different definitions over the years since it rose through the American lexicon, alt-right is mostly used to refer to a movement composed of disaffected young white men who are motivated by fears that whiteness, as a concept or distinct culture, is being eliminated or replaced with the "other." Other, of course being either a nonwhite ethnicity or nationality or, more often, both.

The characteristics associated with the term are mostly correctly applied to those who fit it. They are anti-Semitic and resentful of Israel's close friendship with the United States; they are virulently opposed to illegal immigration, current legal immigration policies, and the ideas that they blanket identify as "amnesty" or "open borders."

These individuals pepper their speech and social media accounts with deliberately shocking rhetoric that proposes that any immigration from nonwhite countries is contributing to the dilution of the theoretically endangered "white culture." They often use the term *white genocide* as a summary of this view.

While their views might be considered fairly standard for any white supremacist ideology or group, the alt-right is more specifically famous for "trolling" online.

Trolling, as you are no doubt sadly aware, is the Internet practice of being a repulsive person. It is often done under the rubric of "moving the Overton window," a reference to the range of ideas that are acceptable in public discourse. In other words, trolls say something utterly vile to move the window further toward the freedom to be vile and further away from the controls necessary to prevent vileness. They claim they've extended the boundaries of free speech. Needless to say, this excuse is often a thin cover for what is actually just a deep desire to hate the Other.

I was first exposed to these alt-right trolls on Twitter, where my stated opposition to Donald Trump's candidacy earned me status as a "cuck." If you are a very lucky and blessed person, you may be unfamiliar with this term, but it was and still is widely and freely used in these circles. In the last year or so, the term has become increasingly acceptable even outside of those circles, used by conservative pundits and writers who once found the term grotesque.

"Cuck" is short for "cuckold," which is defined as "a man whose wife is sexually unfaithful, often regarded as an object of derision."

Eventually, as it was being used primarily to deride conservatives who did not support Trump, the words merged and became "cuckservative." But whether "cuck" or "cuckservative," the meaning is the same.

Several left-leaning websites wrote articles defining it during the election year, but I found *Salon* to provide the most accurate definition.

> *"Cuckservative," you see, is short for a cuckolded conservative. It's not about a Republican whose wife is cheating on him, but one whose country is being taken away from him, and who's too cowardly to do anything about it.*
>
> *OK, that's gross and sexist enough already, but there's more. It apparently comes from a kind of pornography known as "cuck," in which a white husband, either in shame or lust, watches his wife be taken by a black man. Lewis explains it this way: "A cuckservative is, therefore, a race traitor."*[2]

These aspects of the alt-right were clear to anyone who spent any amount of time engaging in political discourse on social media.

Normally anonymous and primarily engaged with online, it was not uncommon to have violent imagery depicting Jewish subservience toward any conservative that objected to Trump.

At *National Review*, columnist David French wrote about the types of "trolling" he received at the hands of the alt-right.

> *I saw images of my daughter's face in gas chambers, with a smiling Trump in a Nazi uniform preparing to press a button and kill her. I saw her face photo-shopped into images of slaves. She was called a "niglet" and a "dindu." The alt-right unleashed on my wife, Nancy, claiming that she had slept with black men while I was deployed to Iraq, and that I loved to watch while she had sex with "black bucks." People sent her pornographic images of black men having sex with white women, with someone photoshopped to look like me, watching.*[3]

At first, there remained what seemed to be a universal disgust of the alt-right. But the alt-right had a lot in common with the run-of-the-mill conservatives who would ultimately elect Donald Trump, at least inasmuch as they both preferred certain policy objectives. One of the easiest ways to reconcile commonalities was naturally the shared disgust with political correctness.

Like most on the right, the alt-right also viewed political correctness as a form of dangerous censorship that would eventually, if left unchecked, open the door to suppression of freedom of speech.

But for the alt-right, the "truths" that were being silenced were the belief that Jews headed a global cabal, that immigration leads to "white genocide," and that black people the world over are inferior, both culturally and intellectually. Their responses to being attacked for these ideas boiled down to this: "It may not be nice, but is it wrong?" Usually combined with some sort of offensive language or photo to indicate how much joy they took in seeing people offended by words.

Fashioning themselves as bold warriors of uncomfortable truths, they helped morph ideas that had been central to conservative thinking into something almost exclusively designed to invoke offense in their opponents. Seemingly for the sole intent of relishing the cuck and liberal tears.

So, where did the alt-right go? In the years since the election, one could be forgiven for believing their power and influence is diminished. The *new* new right, which is to say the new or advanced Trump right, is quick to characterize the alt-right as a fringe element that has no influence on the GOP and for which the numbers have been inflated and exaggerated. While it may be true of individual accounts or websites or groups, what is more broadly the case is simply that the online right has adopted many of the very tactics, namely trolling and vice signaling, that made the alt-right famous in the first place.

There are those on the left who would say the idea that the true conservative right was "hijacked" is overly generous, and that the ingredients of the alt-right had always been integral to the Republican worldview, but that misses the significance of the moment. Whether the components had always existed in silence or had been a development of only the last several years is irrelevant to the argument at hand. The relevant fact is that the alt-right *was* able to seize the Trump moment in time and, through his candidacy and election, provide a framework for what would eventually become an operating tactical philosophy for the movement at large.

A movement, I hasten to add, that is composed in no small part of self-identified Christians.

It's really impossible to say if this was a confluence of events or a wicked and ingenious design on the part of Trump, but the outcome was certainly a convergence. Consider the factors of the moment: Conservatives were done playing nice with their enemies. They were past the point of caring about anyone's sob stories. They were intent on excusing any lack of empathy as allegiance to cold hard facts, and they were quite ready to Make America Great Again, through any means necessary.

Conditions were ideal, and the marriage was consummated. The alt-right, the Trump right, it was just the right, now.

What they needed at this point were guidelines. A handy reference to a new way of thinking.

In truth, they were offered several sets. More than one statement of principles or set of ideas. For the new right, it was an exciting time to have Machiavellian ideas. And rules. A different set of rules to live by.

The new rules.

Revenge

We should pause here to recap a bit. Taking stock, we have an evangelical movement that has gone Machiavellian; we have replaced the compassionate conservative with the bitter one; and the alt-right has integrated itself with the mainstream right. Everyone is wearing a Trump outfit, and nobody wants to give toys to foreign children.

Now it is time for the culture war to enter the conquest phase.

Step one is the low road to enlightenment. As flawed and fallen human beings, the attraction of stomping through filth toward a supposed greater truth is undeniable as a temptation. It is a belief that only through debasement can one reach enlightenment.

There is this notion, especially prevalent on the right and involving no small amount of self-deception, that fighting fire with fire (the idea that you deal with threats by adopting their methods) means "hitting back twice as hard" or going "scorched Earth."

If that sounds like mixing metaphors, that's because it is. Meeting your enemy with an equal response and comparable tactics is not the same as total annihilation. But to the new right, the idea is that, in order to prove to a liberal that their characterization of motive for an event is uncalled for or unjust, one must be orders of magnitude more hyperbolic and unforgiving in reply.

Self-righteous bitterness isn't an entirely unfounded concept. As someone who has indulged, I can say I still believe that accusations of political incorrectness are overused in circumstances where partisans wish to change the substance of a debate and create a strawman of their opponent's motivations. And rejection of that is a natural response.

But there lies poison, and to demonstrate and discuss it, I will once again share a personal anecdote. We go back now to 2014.

Late in the year, the police shooting of Ferguson, Missouri, resident Michael Brown by police officer Darren Wilson, was consuming the news cycle.

Early reports that Brown, a young black man, had surrendered prior to the shooting by the white police officer, had caused protests and at times riots at the apparent injustice. In truth, my first reaction based on these reports was outrage. I have always had what I consider to be a healthy skepticism of power, and in many cases, the police represent a power that should always be watched closely. So initially, I offered a tempered caution mixed with anger that such an injustice might have happened.

As the reports from the investigation came out and they found no evidence of the surrender (and, in fact, concluded that Wilson's shooting was justified and that they would not pursue charges), the mood in Ferguson quickly turned into rage, and the streets were soon awash with violence.

At the time, I was appearing nightly on a television panel where we discussed these issues, and I had been becoming more and more incensed that it seemed no amount of evidence in this case would impact or dissuade those who had decided the issue was racism. The presumption of guilt by a white police officer was, to me, endemic of the politically correct warfare I perceived to be waged on conservatives as well as white people in general.

I'd just found out about an elderly man who'd been beaten with his own oxygen supply and then had his car stolen. On the amateur video of the event, I heard one of the bystanders respond to it by saying, "They made a mistake by giving us no justice."

Enraged at the words I'd just heard on the video, I gave in to my own frustration at the situation and took to Twitter to vent that outrage. I did not consider even for a moment whether or not the unknown person speaking in the video was in any way connected to the movement of "thought police" I abhorred.

No, I had a blunt, unapologetic message that for some reason I believed absolutely had to be spoken to my thousands of Twitter followers. So I angrily announced, "Give me a gun. Put me in Darren Wilson's shoes. I'd have shot Mike Brown right in his face."

The criticism I expected, and I must admit some part of me hoped for, was quick and brutal.

Of my tweet, *Blue Nation Review* blogger Jesse Berney said, "Not only does Howe think Wilson is justified, but he fantasizes himself about shooting an unarmed teenager in the face. If you don't think that has anything to do with the color of his skin, then you're just willfully blind to what's wrong in this country." He added, "I guess this tweet qualifies as dumb, but mostly it seems inhuman."

The tweet was also highlighted at highly trafficked left-leaning websites such as *Talking Points Memo* and *The Inquisitr*, as well as being referenced by prominent liberal pundits and journalists across the blogosphere.

From my perspective at the time, they were proving my point. As I saw it, my detractors were incapable of confronting my tweet at face value without inferring motives that I did not believe were evident at all from the content.

Eventually, I took to *RedState* and offered a defense, without apology, of what I'd said. I attempted to offer a cold look at the facts and to defend my words as some sort of shocking way to draw attention to those facts. It was not that I didn't have empathy for Brown, it was that I thought, if my life were in danger, I'd choose his death over mine, and that I shouldn't have to apologize for feeling that way.

I considered this more nuanced position to be a fairly well-thought-out explanation for the brief and hyperbolic thought I'd put into the ethos. But the truth is, I wasn't really addressing the folly I'd committed. My belief that Wilson was justified, while it may disgust some who still believe it was not, is at least a defensible position. But much like the alt-right, rather than offering a reasonable position, I drew a picture for my audience that sounded like nothing short of pleasure.

In my bitterness over racial identity politics, I'd spoken what I believed was an uncomfortable truth. I'd explained the motivation of that truth, and I'd highlighted the hypocrisy of those who chose to project racist motivations onto me that I did not intend or harbor. But what I hadn't considered at the time I tweeted it was that it was not only Darren Wilson or Michael Brown into whose shoes I needed to put myself.

Outside of those two main characters in this tragedy were

family members. Outside of Darren Wilson and Michael Brown were observers: children, the elderly, people who were perhaps uncertain how to feel about what had happened, and others who were all but decided.

My tweet did not help anyone get closer to a mutual understanding of the issues. It did not assist in making the case that Wilson was justified, nor did it hinder any attempts in making the case that Brown was a victim. It served only to be outrageous and provocative, a way for me to personally feel better about my own frustrations at what had happened to the elderly victim of assault and theft in Ferguson while baiting those I felt certain would apply racial motivations to my words.

I had responded with self-righteous bitterness. If I'm being *really* honest, I said something I knew would or could be hurtful, specifically for the purpose of being hurtful, because when you're angry, punching things feels good.

In the years following some personal but life-altering events, I stumbled into a period of self-reflection that was painful and inescapable. And upon reflecting on that moment in November 2014, it was hard for me to deny that this had been most of how I had conducted myself for a very long time.

I had become so self-righteously bitter about how people dealt with uncomfortable facts that I spent significant time and great effort stoking the flames of emotion for no reason other than to goad the reaction out of my "enemy." An enemy I believed I could use to attract attention to the "feelings over facts" mentality that I deplored. I didn't know it at the time, but I was employing what would become known as the "new rules."

It's difficult to overstate how perfectly this demonstrates the mentality that found glorious purpose in the rise of Donald Trump.

Conservative author and columnist Ret. Col. Kurt Schlichter

may not have coined the term *new rules*, but he is one of its most devoted evangelists. Citing author and radical leftist provocateur Saul Alinsky's 1971 book, *Rules for Radicals*, Schlichter and others contend that conservatives need to apply Alinsky's guidelines for social change.

Written essentially as a sort of treatise on how conservatives should defend the freshly inaugurated President Trump, Schlichter laid out the "new rules" in a 2017 *Townhall* column. While entertainingly written with plenty of tongue-in-cheek style, his rules are nevertheless meant and taken seriously, and I would argue they represent a widely accepted and practiced point of view among today's conservative activists as an answer to years of politically correct targeting of the right.

Among the twelve rules listed, rule four stands out as what I would argue is an abdication of principle in the service of winning.

> Rule 4: *"Make the enemy live up to its own book of rules."* This
> is not so much about pointing out the lies and hypocrisy that
> constitute Leftist orthodoxy—the vicious racism they deny is
> racism because it's anti-white, the racism against non-whites who
> refuse to serve a liberal master, the sexism against women who
> think babies should actually be born, and so on. It's about not
> letting them tie us into knots by using our morals and values as
> bear traps to immobilize and neutralize us. Fortunately, most of us
> have discovered how losing our superficial "political values" helps
> us regain our freedom. We have embraced the power of not #caring.
> And liberals have no idea what to do when they shout "Trump is
> a meanie," and we shrug, smile, and bust out with an impromptu
> interpretive dance to celebrate Neil Gorsuch.

What is being suggested is that in our weakness in the face of indecency, we either were bound to standards of decency we'd constructed ourselves but whose meaning we no longer understood, or had inexplicably accepted standards of decency our adversaries had constructed for us.

We handicapped conservativism by accepting the limitations required to take the "high road" and follow the rules of polite, "politically correct" society. Meanwhile, our adversaries, who often created those very rules of decency, were not bound to any such expectations, thus making our losses inevitable.

Schichter argues that these standards should not be a millstone around our necks, preventing us from achieving the desired outcome. At the very least these standards should be relative to the objective and considerate of the opposition. In other words, we cannot be so beholden to the defense of a value that we undermine the intended outcome of that value.

If I'm so committed to nonviolence that I do not intervene as a man is beaten to death in front of me, I haven't really protected that value so much as I've protected my superficial projection of it.

Furthermore, by Schlichter's telling, the left had no such decency in the first place, even as they were free to box conservatives in by demanding they uphold a standard the left was not required to uphold themselves.

What noble purpose is there, he seems to wonder, in being the most decent dead man on the battlefield?

This philosophy is perhaps not uncommon when faced with similar situations one might experience in everyday life.

If you are continually finding yourself at odds with a neighbor who is unconcerned about disturbing you when he blasts loud music late at night (he sleeps during the day), an easy and very human reaction on your part is to play loud music during the day until your neighbor grasps how rude it is. Never mind that you yourself had to become rude to accomplish this.

Putting aside that this doesn't take into consideration any other of your neighbors whom you've now disturbed in your pursuit of parity, if you believed your neighbor's actions were indecent, you've now rationalized that indecency itself is your only avenue to peace.

Alinsky and Schlichter employ this rationale with a singular objective that applies to both politics and the neighbor analogy. Winning.

Perhaps you will win a small victory (the neighbor stops blasting music) or perhaps you win something much larger (the neighbor moves), but in either case you've achieved the end you sought, and the means, regardless of what particular values you may have shed in your effort, are a small loss when finally enjoying a good night's sleep.

Rule five is also one of the more widespread in current conservative thinking.

> Rule 5: "Ridicule is man's most potent weapon." Actually, the AR15 is a more potent weapon, but ridicule will do as long as the Left doesn't try to make good on its countless threats of violence and tyranny. Regardless, we finally have a conservative corps that is willing to mock the members of that motley collection of pompous, inept, lying jerks we call the Democrat Party and its media catamite corps. When they turn around and try to mock us back, well, we aren't watching their late night hack comics anymore, and frankly they can make all the jokes they want. The punchline is still going to be "And then the Republicans repealed Obamacare."

In addition to the exhaustion conservatives felt as a result of the impugning of their motives, political correctness had long been derided as an oversensitivity to objective truth. It was often seen as a way to simply deny that which was uncomfortable.

Schlichter's use of rule five is his answer to this problem. It's the metaphorical slapping of a panicked person to shock them back toward reality. Conservatives feel that Americans have become too coddled. That the idea of calm and polite explaining of something difficult to hear may have its place, but that the pendulum has swung too far in favor of feelings.

I can't speak to whether these philosophies might function well in a universe that is devoid of absolute morality, but that is not the universe I occupy, nor is it the universe that Christians should believe they occupy.

Schlichter's strategy means winning a battle but losing a war. His argument presumes, of course, that this is a war, and he employs hyperbolic martial imagery to whip up the base and further the false impression that this is a zero-sum game. But even taking his premise at face value, as Christians we cannot accept an amoral view of war. Winning at any cost is not worth it. It's not biblical, either.

For the "new rules" to work, one must believe that morality is relative to the situation. That, in light of more pressing concerns, we must forgo such fanciful notions as "decency" because the stakes are simply too high.

It's all just moral relativism, which for some may be just fine, but for Christians should be a much bigger problem.

The first and most obvious problem Christians must contend with when attempting to employ moral relativism is that we are never tasked with determining morality by comparing people to *other* people. Christians are tasked with determining decisions of morality by comparing themselves to *Christ*.

The earliest example of how God shows us that these determinations are not made by looking inward or around but by looking upward is in Genesis.

In Genesis, chapter 1, verse 4, "God saw that the light was good." In verse 10, of the creation of land and seas, the Bible also says, "God saw that it was good." The same observation by God is noted about the creation of food and vegetation, about the separation of night and day, about the creation of animals, and about Adam himself, where for the first time among the seven instances in chapter 1 it specifies of His creation that God saw it was "very" good.

Additionally, of the tree that contained the forbidden fruit, God refers to it as "the tree of the knowledge of good and evil." But notice that when Eve first decides to eat of this fruit, it reads, "When the woman saw that the fruit of the tree was good for food and pleasing to the eye, and also desirable for gaining wisdom, she took some and ate it."

So why so much focus on this word in the early chapters of Genesis? What is the significance? For that matter, why put a tree into a garden for the express purpose of forbidding Adam and Eve from eating its fruit?

Of all the reasons for God to place the tree in the garden or use the word *good* so often, the one that stands out to me the most is a common theme. He simply intended for a choice to occur. The tree itself did not contain pixie dust that provided knowledge of good and evil. But the choice itself, once made, becomes that knowledge. By eating the fruit, and determining for themselves what was good or evil, Adam and Eve had instant knowledge of the nature of that choice. And with it, the deadly cost of looking inward, instead of upward, for determining right and wrong.

Put more succinctly, they rationalized to themselves why what they could attain was in their self-interest, and they chose to believe whatever they must to justify disobeying God.

In the immediate, they became self-aware of their ability to choose for themselves. In the long term, they doomed the world. Their regret must have been potent in the wake of this decision, but throughout the Bible, God reiterates the importance of the lesson that we do not look to ourselves to establish moral certainty.

Proverbs 14:12–15 drives this home:

> [12] There is a way that appears to be right,
> but in the end it leads to death.
> [13] Even in laughter the heart may ache,
> and rejoicing may end in grief.
> [14] The faithless will be fully repaid for their ways,
> and the good rewarded for theirs.
> [15] The simple believe anything,
> but the prudent give thought to their steps.

In all things there must be wisdom, and often it comes on the tail end of the warnings above. The same will be true of conservatives and evangelicals who, in their rush to protect their own vision of desirable ends, embrace *any* means and tell themselves there will be no cost.

The "new rules," if employed by Christians, are an abomination of God's promise—the promise that trust in Him through the appropriate means will align us with the ends He has designed.

Instead, the "new rules" rest in the notion that God needs us so badly to achieve His ends that we must be willing to sacrifice anything, even His very commands, in order to accomplish His will.

This was an easy deception to adopt as faith began to shift away from the morality and empathy of a graceful and merciful God and instead move toward an absolute allegiance to truth as an offensive weapon. Truth is sometimes hard to hear, but that difficulty is no excuse for denying its existence. Part of growing up is learning this, and the right seems to feel that it is long past time for the left to slide into adulthood.

If the pendulum had indeed swung too far in favor of this coddling, the right seemed to embrace the "new rules" with an enthusiasm that merely swung the other way, perhaps by design. The idea, again, of fighting fire with fire. The left went too far in one direction, so to bring them back to the center we must go equally as far in the other.

This is exemplified with the mantra of author and *Daily Wire* editor in chief Ben Shapiro: "Facts don't care about your feelings." It is a clever summary of what is a fairly normal and frankly uncontroversial conservative view. The premise is obviously that regardless of how uncomfortable the truth may be, the truth is the truth. But one could be forgiven for thinking many on the right seem to subscribe to something closer to "facts are here *for the purpose of hurting your feelings.*"

The late Andrew Breitbart, who was a towering figure in politically conservative citizen journalism, famously said, "Truth isn't mean, it's truth." And again, the idea seems to have morphed. At first to, "Truth isn't mean; it's a weapon." At times, even, "If it's not mean, it's *not* the truth." This idea fits the GOP in the Trump era much better and is arguably the way many conservatives have chosen to employ the "new rules."

Political correctness as a suppressive weapon against good faith disagreement persists, and that is, of course, bad, but meaningful opposition to it has seemingly been replaced with what amounts to its own form of verbal tyranny. Also, bad. So it is that any appeal to decency or even common courtesy is decried as virtue signaling from today's right.

When the *Washington Post* reported that President Trump said in reference to immigration from poor African nations, "Why are we having all these people from shithole countries come here?" the condemnation was swift from the left and from portions of the right as well.

But among advocates of the "new rules," even prior to Trump's eventual denial, while some not only defended the verbiage and attacked any rebuke of it as virtue signaling, others softly objected to the language while still defending its implication.

Robert Jeffress told the Christian Broadcasting Network that "apart from the vocabulary attributed to him, President Trump is right on target in his sentiment. As individual Christians, we have a biblical responsibility to place the needs of others above our own, but as commander in chief, President Trump has the constitutional responsibility to place the interest of our nation above the needs of other countries," adding, "I'm grateful we have a president like Donald Trump who understands that distinction and has the courage to protect the well-being of our nation."

It is hard not to translate this as Jeffress suggesting that Trump is selfish on our behalf so we can continue to pretend to put the needs of others before ourselves. It is, in other words, one more rationalization by the Christian right in embrace of worldly values and—I must employ the term again—a Machiavellian worldview.

Victimization

The "shithole" comment had legs. It generated a lot of content for the cable news shows, the columnists, and the bloggers. That's a sort of specialty of Donald Trump's.

Fox News host Tucker Carlson defended Trump's language, as

well as his intent, saying, "So, if you say Norway is a better place to live, and Haiti is kind of a hole, well, anyone who's been to those countries or has lived in them would agree." He added, "But we're jumping up and down, 'Oh, you can't say that.' Why can't you say that?"

Conservative pundit Ann Coulter tweeted, "Okay, yes—Trump shouldn't call them 'shithole countries.' A little respect is in order. They are shithole nations."

Tomi Lahren, a contributor to *Hannity on Fox News*, tweeted, "If they aren't shithole countries, why don't their citizens stay there? Let's be honest. Call it like it is."

Jack Posobiec, primarily recognized as someone who loudly supports Donald Trump, also weighed in, tweeting, "Its remarkably refreshing to hear a President who speaks his mind instead of lying to the American people every day."

Most of the criticisms were treated by the right as "fake news." Often, critics were told to "move to Haiti if it is so great."

When Mitt Romney said on Twitter, "The poverty of an aspiring immigrant's nation of origin is as irrelevant as their race. The sentiment attributed to POTUS is inconsistent w/ America's history and antithetical to American values. May our memory of Dr. King buoy our hope for unity, greatness, & 'charity for all,'" Chris Pandolfo reacted at Conservative Review, saying, "It looks like Romney will be focused on virtue-signaling."

Rush Limbaugh, radio host and one of the most influential conservatives in American politics, said on his radio broadcast of the remarks, "I have been in the midst of these kinds of firestorms, folks. And I can tell you this is all faux rage. It is faux anger. It is faux outrage. It is made up. It is for the cameras. It's for the microphones. It's for the audience."

These reactions are a perfect display of the importance that Schlichter's fifth rule plays in modern conservative politics.

In other contexts or in previous presidencies, such a remark might've been met with condemnation from any or all of these same figures. But to conservatives, the neighbor playing the loud music for so long had already enjoyed too little in the way of opposition. He'd been enabled for far too long and had grown entitled to a behavior that harmed others. If the other neighbors would not stand up to him, someone must.

Conservatives saw the injustice of the maligning of their motives and character for years. They were called racist and sexist for so long and in so many instances where it defied reason that such a motive could exist, their hearts had hardened. Far beyond simply becoming desensitized to the accusation, they had become instinctually skeptical of its existence.

"Offense" had become so completely irremovable from political posturing that, to many, there was little point in parsing anymore. If an accusation is directed toward anyone deemed valuable to the right's interests, that accusation will be denied as invalid. Any detractors on the right who see it differently will be accused of virtue signaling.

A perfect protective shell. In fact, one of the most important parts of Trumpism is this shell, even if it's not so clearly stated. Even for those who aren't activists, or who aren't engaged online. You can hear similar sentiments from true Trump believers anywhere in the country, spoken in a million ways. Just ask Trump voters what they think about people being offended by things. Ask about the Washington Redskins or quotas or diversity or any of the trigger words, and you'll get the same idea.

Everyone who is offended is on the left, and everyone who wishes not to offend is a tool of the left. No matter how folksy the phrasing, you'll hear it. Guaranteed.

From the long-ago sentiment "don't tread on me" had evolved a new way of thinking. That ubiquitous, pervasive point of view has even led to new language. In the movement, and particularly in the most public part of it, any and all opponents who take issue with a conservative's words or actions became known as "snowflakes."

What's so oddly circular about all this is, of course, the fact that they are essentially indulging in exactly what they are decrying. They are breaking the old rule by following the new rule. And the rule is good because, if you follow the rule, then it must be good.

So now in place of debate you have argument, and within each argument you have both sides, right and left, always assuming bad faith while also acting in it. And attempting to silence opposition through designation rather than engagement. *I'm a racist? Well, you're a snowflake!*

It's the triumph of the paranoid style.

If you don't buy it, try it. Take any politically correct accusation abhorred by the right and replace "sexist" or "racist" with "snowflake" or "Trump derangement syndrome" and you will hardly be able to tell the difference.

So, utterly disgusted with liberal victimhood politics, conservatives adopted their own.

It's important to remember that this is being undertaken *deliberately* by the opinion makers on the right. The soil may have become fertile on its own, but the planting in that soil has been determined and organized. It was new to the right. It required guidance.

That newness made its appearance a novelty, and so a lot of people took note.

In a column in the *Atlantic* titled "A Nation of Snowflakes," writer Adam Serwer noted this growing victim mentality on the right, saying, "Even as they portray liberals and leftists as weak snowflakes, conservative complaints about political correctness often reflect acute sensitivity to liberal or left-wing criticism—criticism that when they can, they try to silence through opprobrium."

In other words, conservatives tend to portray the maligning of their motives as just as deeply hurtful as a liberal might portray the assumption of ethnic or gender inferiority. It's another very important aspect of this new right. Being unjustly called racist was an equal injustice to being the victim of racism. Thus, they could be utterly outraged and therefore justified in rejecting critical voices from outsiders.

This silencing was done in the exact same way they had perceived it as being done to them. Their own version of political correctness had been forged, a set of silencing rules just as strict as those of the P.C. police, and it was viewed as merely being honest. Facts, not feelings, and so on.

So the philosophy at work here has three assumptions: First, that the persecution of the right is real, ongoing, and must be stopped. Second, that being silenced through accusations or character assassination would never happen to them again, but that they would make sure it happened to those who opposed. And third, that the idea of offense was itself so offensive that the only cure was to be deliberately and provocatively offensive. "Truthful," they call it.

The third aspect was the most vital. It was the one that satisfied vengeance, that pushed the limits. It moved the Overton window and acted as a shield against being called names. Therefore, it was the most important thing to adopt. "Brutal honesty," as they believed it to be, became a nearly religious value to the Church of Winning. The virtue of the virtueless. And what's more, the means to an end.

Merging this deeply felt state of being with Donald Trump's Breitbart/Bannon–allied candidacy was a completely natural and inevitable act. He stood head and shoulders above anyone else in politics as someone who not only held the view that the era of "political correctness" was at an end, but who in practice embodied the most important ideals of the "brutal truth" philosophy.

Trump's speech about Mexico and Mexicans was carefully crafted and worded. It was meant to do two things, and it did them to utter perfection.

First, it was meant as exactly the dog whistle people said that it was. When he talked about Mexican rapists, that nationalist and even racist impulse on the right heard him loud and clear. This is something he continued to do throughout the primary, and it worked like a charm. The racists, whom he would reject out loud, knew what he was "really" saying, and they were utterly loyal to him. In most ways, they still are.

Second, it was meant to make liberals say, "This is racist," which they did. Which set him and the movement up perfectly to parse the actual words. "He said they send their rapists, not that Mexicans are rapists." And that was true; that *is* what he said. And saying that is defensible. That was a sweet song to all conservatives. The immediate reaction was to defend the speech, to attack the liberals for crying racist when what he said was perfectly fair, and so on.

That put Trump on their side and, much more important, them on his. The right was defending him right out of the gate. Us against them. He and we versus you and they. Any Republicans, including opponents, who said anything negative about his speech put themselves in opposition to the angry base. That was the majority of the base. Angry about political correctness, angry about being called racists. Angry they couldn't be practical or pragmatic . . . all of it.

And there was Trump, saying what everyone else was afraid to say. Not cowering before the language police.

Even more than that, though, everyone knew and believed he did it on purpose to pull the strings of the left and media, and they liked that even more. Trump was not merely unafraid of giving offense, he actively gave offense. On purpose. Merrily, even. He was the thumb in the eye. He was the guy at the party not afraid to tell what is "just a joke." He was the "get over it" candidate.

Trump was practically a saint of the new church before he ever won a primary.

Nevertheless, he did still have to *win* those primaries. He still had to get from being a figurehead in September to his eventual place as de facto leader of, among other groups, evangelicals in America.

That meant overcoming a few shared obstacles.

Trump had to move to the forefront of not just the election but the center of political life for evangelical voters. That meant not only tapping this mentality but exemplifying it.

The religious right, rather than learning from experience, became bitter in their losses. Through that bitterness they embraced a mentality of victimization. They believed they were not the architects of their defeat, but rather that the rules and the game had been stacked against them.

This is one of the most important features of Trumpism. The victim American.

Christians, conservatives, evangelicals—the base voter for the right wing, in other words—in order to reverse what they saw as something that had been done to them must fight fire with fire. The rules must change. Turnabout became fair play, and almost the entirety of the conservative movement became a feeding frenzy of outrage peddling and persecution complexes.

No longer tasked with showing a *better* way, these groups now had a glorious obligation to force the hand of opponents through scorn and malice, allowing and participating in whatever lies or deceit were necessary, in service to this greater good that was, theoretically, a shared goal and one the right believed could be achieved.

This sort of disease of thinking that the movement was displaying is an old one, and I'd wager it's one that has infected far more than just Trump's particular base. It is a disease of delusion, one common among ultra-partisans, but that, theoretically, God-fearing Christians should be especially guarded against.

Normally, or at least in my own experience, this sort of self-delusion is in service of believing a narrative, and often that narrative exists as a way of making sense of the world.

I have very real experience in this.

When I watched Donald Trump's rise and saw the changing standards and the betrayals of underlying principles among those I had called my allies for so long, it quickly became difficult for me to deny the truth.

After spending so many years swearing that our underlying principles, and for many of us our devotion to the sovereignty of God, was our driving force, I saw ally after ally claiming that Trump's victory in the primaries left us no choice. We were completely deprived of choice. Only one course of action could be right. We had to support Trump, and to put any and all of our reservations aside in deference to the *ends* that he was especially or particularly equipped to achieve, or at least more prepared to achieve than the alternatives.

In May 2016, after Ted Cruz withdrew and Trump's nomination was all but secured, I laid out in stark terms at *RedState* the choice I felt was in front of me: a reckoning of sorts. It was one that found me coming face-to-face with all of the things I'd looked past as well as with the objectively bad things I'd ignored, supposedly in service of a greater good I'd decided was paramount. I laid out in detail how it had all blown up in my face.

It was one of many articles I wrote that year attempting to articulate the belief that service of a greater good cannot be achieved with a lesser means. At least not in any way that I've found the Bible to support.

"I chose peace over principle," I wrote. "I chose to go along with those I disagreed with on core matters because I believed we were jointly fighting for other things that were more important. I ignored my gut and my moral compass."

No longer mincing my words, I concluded by skewering my former allies in conservatism. "They are in this for money and power and influence and they think Donald Trump is their ticket," I wrote. "Hell, they may be right. And I'll go down in flames with my principles before I join them."

The article garnered significant attention across the Web and on television at the time, but it turned out to be little more than a blip on the screen to the intended audience. The broader spectrum of noise demanding that all Republicans "get behind Trump" for "the good of the country" was far too overwhelming.

And even among those to whom it made a difference and who discussed and considered this point, it was in large part rejected whole cloth. The tide was set. There was little that would resist it.

Selfishness

Even two years after the election, experts are still trying to put together what precisely caused evangelicals to vote in record numbers for someone so seemingly adverse to the notion of Christian values.

The answer was given over and over throughout his candidacy and in the immediate aftermath of his election by many reluctant supporters who regularly referred to Trump as "not their first choice." Quite often, if not mainly, they claimed that the very values they'd abandoned were the ones that had instructed their decision.

In Wayne Grudem's retracted (and later reinstated) column "Why Voting for Donald Trump Is a Morally Good Choice," he specifies from a value-centric perspective why Trump gets a morality pass:

> *This year we have an unusual opportunity to defeat Hillary Clinton and the pro-abortion, pro-gender-confusion, anti-religious liberty, tax-and-spend, big government liberalism that she champions. I believe that defeating that kind of liberalism would be a morally right action. Therefore I feel the force of the words of James: "Whoever knows the right thing to do and fails to do it, for him it is sin" (James 4:17).*[4]

Prominent evangelical leaders weren't nearly as cautious in their tone, with most making the case that God's purposes should not only be the focus of voters, but that they are the focus of Donald Trump as well. A vote for Trump *is* a vote for Christ's values.

In an interview with *Christianity Today*, Dr. James Dobson discussed his reasons for endorsing Trump, saying, "I liked that he promised us emphatically that he will work to protect our religious liberties. He has since released a list of potential Supreme Court nominees that is stellar. We must pray that, if elected, he will keep his word."[5]

Prosperity gospel advocate and member of Trump's Evangelical Advisory Board pastor Paula White, told the *Christian Post*,

"People have a clear choice this election: a pro-life Donald Trump or a pro-choice Hillary Clinton."

Another Evangelical Advisory Board member, James Robison, went further in offering the Christ-centric version of Trump advocacy.

> Robison said that Trump could potentially be "the people's champion," like a Saul who "is going to let David, the shepherd, in the room."
>
> "As a matter of fact, I think there is a great likelihood that he will invite the shepherd that David represented in the room," Robison stated, saying that the times he has spent in prayer with Trump has [sic] been "as special as anything I have ever seen."
>
> "I believe [God] is offering the most powerful, grace-filled invitation in history to this nation to come back to the shelter and shadow of the Father and the watchcare of the shepherd," Robison stated. "He is wanting us to come like chicks to the wings of a hen. I think the stage is set for us to come and I honestly believe that [the Trump] family will actually seek that wisdom."[6]

It's worth noting that the man who "let David, the shepherd, in the room" later tried to kill him. It was a famously *false* friendship.

Given so much focus by these leaders on how the motivations of a Trump vote must consider the weight of Christian values, they might be surprised to learn that their various flocks apparently had no such priority.

While the common wisdom seemed to be the issue of abortion specifically, an easily identifiable Christian value, polling told a different story.

The University of Chicago's Divinity School broke down polling from LifeWay Research, which had been conducted in the final months of the election. It found that only 4 percent of evangelicals cited abortion as their determining factor in a candidate.[7]

When asked what issue mattered most, evangelicals, by a wide margin, chose terrorism and the economy (26 percent and 22 percent, respectively). In fact, abortion ranked at the very bottom behind immigration, with religious liberty coming in third to last.

A Pew Research poll showed similar priorities among evangelical supporters of Trump.

Whereas LifeWay Research ranked their poll by having voters choose what was "most important," Pew Research measured simply by what evangelicals selected as "very important" and found similar results.

Terrorism (89 percent) and the economy (87 percent) led the pack again, closely followed by immigration (78 percent).

Abortion (52 percent) ranked fourth from the bottom of fourteen issues, edging out treatment of racial and ethnic minorities (51 percent), the environment (34 percent), and treatment of the LGBT community (29 percent).

While respondents were unable to choose which issues were under consideration beyond the fourteen presented, abortion is arguably the only one most associated with evangelicalism.

And yet it ranked behind foreign policy (78 percent), gun policy (77 percent), Supreme Court appointments (70 percent), health care (70 percent), Social Security (70 percent), trade policy (62 percent), and education (59 percent).[8]

But perhaps more important, what do those specific priorities say about the evangelical movement?

It seems to say that while the reasonable desire to have a prosperous economy and a credible expectation of safety may be fine, for these supporters, a prosperous economy and a credible expectation of safety are far more important than their stated desire to save human life in the womb. By a lot.

The notion of religious liberty is not nearly as important to evangelical voters as the stock market surging or "finally" having a president who's willing to threaten to use nuclear weapons on our adversaries. Whether religious liberty is truly in danger or not is irrelevant to the point.

Evangelicals and conservatives are consistently proposing that abortion is the key issue, and yet it barely falls 4 percentage points below "none of these" as an option on the LifeWay poll.

I could spend the next several pages going on and on about how God does not wish us to live as paupers. That the verse "love of money is at the root of all kinds of evil" is often misquoted and misinterpreted and that it merely indicates money can easily become an idol while not suggesting wealth itself is inherently evil.

But I won't, because the point I'm attempting to direct you toward is that evangelicals purported, for decades, to position the urgent need for the reestablishment of Christian values as the central doctrine of their political motivations. Above all else, we were tasked with growing God's kingdom, preserving His creation, helping the poor, and loving the downtrodden. Despite evangelical leaders' *talk* of character, their followers have the inverse priorities. That these leaders can't recognize that it's their hypocritical actions which have led to this gap between abstract ideals and real-life priorities is precisely reflective of how they've chosen to misuse the mantle of leadership.

By directly defying their stated desire, ignoring the character of Donald Trump, and creating a "Christian" culture that has become divisively self-interested and bitterly self-righteous, these leaders have taught their flocks to value the things of the world, rather than the things of Christ.

After all has been said and done, and more than one hundred pages into this book, it all starts to look so simplistic, doesn't it?

We discussed in the opening chapters how members of the movement had come to contort themselves in defense of what had for so long been considered indefensible.

We subsequently discussed how that shift defied the stated values of a movement that was born in the early twentieth century in the hopes of infusing Christian morality into governance and that led the charge against Bill Clinton with battle cries of "character matters!"

We have seen how the exhaustion of losing, the feelings of persecution, and the seduction of vengeance have animated conservatives and evangelicals into no longer viewing a figure such as Donald Trump as a "lesser evil," but one whose style they were ready to adopt before he even announced his candidacy.

Brutality is the new outreach. Subjugation is the new persuasion. Pragmatism is the new morality. Winning is the new religion.

But all of this failed to answer that final and important question.

Why?

Such a simple answer: selfishness.

Something from which no one is free of temptation but which we are all tasked with rising above.

One of the most remarkable aspects of observing all of this—at first from the inside and then as an outcast apostate following my heretical rejection of his candidacy—is how readily evangelicals emulated some of the most well-documented aspects of Trump's apparent narcissism.

Trump's ego is brittle, despite its immensity, and he obsesses over negative coverage of himself and is seemingly incapable of acknowledging fault. Every loss is a win; every mistake is brilliance.

Trump evangelicals and conservatives en masse have adopted this same posture. Clearly it's working, at least for now.

The question they haven't asked themselves is one they would mock upon hearing: What is the cost of winning?

You've gained the world. How is your soul faring?

Chapter 6

STATE OF
THE CHURCH

T rump's Republican Party did better than many expected it would in the 2018 midterms, traditionally a difficult election for the president's party, but it would be hard to describe it as a victory for Trumpism.

Republicans won twenty of the thirty-six gubernatorial elections, but Democrats unseated twenty-nine Republicans in the House and picked up fourteen open seats, causing the majority to change hands.

However, Republicans had a net gain of two seats in the Senate and maintained the majority of state legislatures with thirty-one Republican-controlled states compared to eighteen Democratic states, with one state having a split legislature.

But if anyone had wondered whether the series of missteps, scandals, firings, and negative news coverage had impacted evangelical voter enthusiasm for Trump, that question has been answered.

Evangelicals overwhelmingly supported Trump's Republican Party, voting for GOP candidates at a margin of 75 percent, a slight dip from the 81 percent that went for Trump in 2016, but dips are expected in midterms.[1]

And that's not from a smaller sampling of voters either, as once again, and for four elections running, evangelicals composed 26 percent of the voting electorate.[2]

But hidden within those numbers is another story.

There's every reason to believe that evangelical enthusiasm is even higher than it appears.

This is because while they have maintained a 26 percent hold on the total voting electorate for several years running, their percentage compared to total population is going down.

Robert P. Jones of Public Religion Research Institute pointed this out on Twitter, noting that evangelicals as a percentage of population has dropped steadily from 20 percent in 2012 down to 15 percent in 2018.

There are several conclusions we can draw from this, but the two primary ones are these:

1. Evangelicalism is losing members.
2. As evangelicalism loses members, they double down on their current path, increasing their enthusiasm to maintain higher and higher turnout so as to keep up at 26 percent of the voting electorate, despite having fewer evangelicals overall from which to draw.

If that sounds like isolation, then I agree. And a movement that can exist only through outreach that chooses instead to close in on itself is doomed.

To make matters worse, it is not only the evangelical sector of Christianity that is losing members. Christian faith overall is declining.

Pew Research reported an 8 percent drop, to 71 percent, of Americans who describe themselves as Christians over the seven-year period between 2008 to 2015.

There is a growing sentiment among Christian leaders that there's no particular need to do anything about this apparent isolation.

Russell Moore, who it should be noted has stood in opposition to Trump, has taken the view that we should just accept the decline of faith overall and hope that as the numbers decline, the stigmas that may keep people away or cause them to leave will regain a sense of strangeness that might attract curiosity.

"It is no longer possible to pretend that we are a Moral Majority," Moore says on the dust jacket of his 2015 book, *Onward: Engaging the Culture Without Losing the Gospel*. "As Christianity seems increasingly strange, and even subversive, to our culture, we have the opportunity to reclaim the freakishness of the gospel, which is what gives it its power in the first place."

Perhaps his theory will prove true, but I would think we should still move forward with attempting to grow the kingdom without counting on the idea that rarity could be the passage to curiosity. To be clear, I'm not suggesting Moore thinks we should encourage a decline in church membership or we should stop preaching the gospel. But his conclusion that losing souls might turn out to be a positive turn of events feels suspiciously like wishful thinking.

Given the nature of Christian values and the fact that answering to something higher than the world is core to its ministry, appearing to abandon that higher calling is more than corrosive to the perception of Christianity among nonbelievers; it is also an opportunity for any who might want to corrupt or replace it.[3]

Far from conspiratorial, the nature of competing ideologies and religions is that they each evangelize and proselytize in service of their beliefs. Any or all of these competing value systems would certainly be happy to move into the abhorred vacuum Christians might leave behind as the culture moves away from them.

One of the key failures American evangelicals have continually been loath to recognize is that the culture doesn't "move away" from them into nothingness. Humans are programmed to seek something to fill their need for understanding everything from the vastness of the universe to their immediate surroundings.

What might replace the Christian ideal? Anything from another faith to secular humanism or even scientology. What will replace it is not nearly as relevant as the urgent need for Christians to wake up to what their continual moaning about the culture will ultimately mean.

Like typically self-centered people, today's evangelicals seem to think of the decline of Christian values as something that's happening *to them*. That the culture is something *they* have lost. Apart from their own decisions to ignore their stated values, it is not Christians who are "losing" something. If you subscribe to the view that Christ is the savior of all mankind, then it is most certainly everyone *else* who is losing something.

Evangelicals have steadily abdicated their cultural leadership role over the decades, choosing instead to depend on government-led crusades or simply closing in on themselves, which has left a vacuum for distortions of Christianity to fill.

Already the polling is available that reflects this decline.

Political polling blog *FiveThirtyEight* reports that "the white evangelical Protestant population in the U.S. has fallen over the past decade, dropping from 23 percent in 2006 to 17 percent in 2016. But equally troubling for those concerned about the vitality of evangelical Christianity, white evangelical Protestants are aging. Today, 62 percent of white evangelical Protestants are at least 50 years old. In 1987, fewer than half (46 percent) were. The median age of white evangelical Protestants today is 55."

The church's demographics are shifting, with more people of color identifying as evangelical than before, but the group's membership still decreasing overall.

Of the reasons offered for the decline, most people cited "belief incompatibility." This can be categorized as any number of things, including perceptions of sexual morality and gender roles, but the solution being offered as a way to repair this is what *FiveThirty-Eight* refers to as "theological flexibility." Essentially, if the population is moving away from Christianity, then the solution must be to conform Christianity to better fit the times.

As a Christian, I must protest.

The issue is not the theology God wishes for us to study and learn. The problem is and continues to be the poor salesmanship by those tasked with making the case for Christ. It is theologically irreconcilable to believe that the nature of God's Word is incompatible with His own creation.

As an illustration of how to fix the problem, consider Apple Computer.

When Steve Jobs triumphantly took the helm back in 1997 after a decade-long absence, Apple was a shambles. In just the three years prior to his return, the company's sales had declined by almost $4 billion.

As part of an attempt to help the company find its footing again, Jobs, the original founder of the company, was brought in to think outside the box and try to rescue the situation, which, of course, meant spending at least some time identifying what was causing the problem.

In a famous presentation at Macworld that year, Jobs spent ample time discussing how fantastic and forward thinking the company's product line was and how much potential there was for new customers to love them.

From leading the world in hardware in service of website design to being the number one computer in education, it was clear Jobs believed the problems Apple faced had much more to do with previous management. But he also identified another key problem for the company, which was the exclusionary culture of the Apple community.

The devoted disciples of Apple had become something of a cult. They had a great fondness for Apple products but generally had a reputation as snobs. They constantly appeared to look down on PC users, mockingly mentioning the superiority of Apple without necessarily going to any lengths to convince people of what made them so great.

In Jobs's presentation, this became most clear when a few audience members booed after learning of Apple's new partnership with competitor Microsoft. After a few additional unhappy voices were heard when Microsoft CEO Bill Gates joined the expo via satellite, the mood of the room appeared mixed.

Once Gates was done talking about the partnership, Jobs addressed the naysayers head-on, saying, "Y'know, where we are right now is we're shepherding some of the greatest assets in the computer industry. And if we want to move forward and see Apple healthy and prospering again we have to let go of a few things. We have to let go of this notion that for Apple to win, Microsoft has to lose. We have to embrace the notion that for Apple to win, Apple has to do a really good job." He continued the apparently improvised remarks, saying, "And if we screw up, it's not somebody else's fault. It's our fault."

This culture shift that Jobs ushered into his company undoubtedly played a role in growing Apple's customer base exponentially. This new culture and leadership catapulted Apple to heights it scarcely may have imagined, with the computer giant's market value now casting it as the most valuable company on Earth with a market capitalization of over $800 billion.

Much like Apple Computer did, at this moment the American Christian Church is facing a crisis. The problem is not what

Christianity has to offer, and it is not Christianity's main competition, the culture at large.

If we want American evangelicalism to be healthy and prospering and growing again, we too have to let go of a few things. We have to let go of the notion that for evangelicalism to succeed, American popular culture has to lose. The reason is simple: American popular culture "losing" is not a guarantee of anything as it relates to Christianity. Without an inclusive and loving Christian culture that seems more intent on inviting people in than wagging our collective finger at them, evangelicalism will lose regardless of popular culture.

We need to own our own mistakes and change our own Christian subculture if we want those who currently see evangelicals as unpersuasive hypocrites ever to be open to listening again. But so far, Christians have not shown an eagerness to rise to that challenge; they seem much more prepared to shed their principles for Donald Trump than they are willing to work to earn the trust of a disenchanted public.

It's difficult to find a better illustration of the lack of prioritization than the fact that the statistics I mentioned above have not even hit the radar of evangelical leaders. Instead, those leaders are happy to devote mountains of time celebrating the return of a seasonal greeting that had never disappeared in the first place.

Far from hoping to fix these issues of Christian culture, evangelicals have seemed content to adopt an offensive position in lieu of any reflection or outreach. That stance comes from years of perceived abuse that has gone unanswered. Abuse from the culture, from the Establishment, from liberals—really from just about the entire world.

But what remains remarkable is that evangelicals, as part of the larger conservative base, have convinced themselves that Trump is the vessel that will lead to that better world. It may be appropriate to think of it less as the "better world" they are claiming to be working toward and more a "better life." Specifically, their own.

I should note that there is nothing inherently wrong, or even religiously wrong, with seeking a better life. Not for one's family nor even for oneself. No, the problem here is the selfishness. When they say "better world," they mean "better life," and they mean it only for their own demographic. That may sound harsh, but the evidence is overwhelming. The new and often stated objective of the political evangelical movement is prosperity for that movement in particular, for renewal in their towns in particular. For their "kind."

You don't hear leaders on the Christian right lamenting the state of schools in the suburbs of Chicago, do you? No, it is always factory towns in the South or mining towns in West Virginia that require aid.

There is a political sense in which one can attribute that to the idea of the missing or left behind, the underrepresented multitude who don't have the benefit of Hollywood charities and liberal favor. In other words, the sort of "you have Beyoncé, we have Dobson" mentality.

But that other aspect exists, too. The tribal one. The insular, isolated, self-identifying one. "Our" communities. The sense of the Christian evangelical as not merely a voting demographic but a society unto itself, and one that requires not only representation but advocacy. To the exclusion of others.

It's a remarkably un-Christian impulse, but one that drives a great deal of commentary on the religious right. And once more, it represents that turn away from God and toward man that is so stark in this ongoing movement.

Polling of pastors, of which many on Trump's Evangelical

Advisory Board can count themselves, showed somewhat different results than those from congregants.

Evangelical pastors chose personal character of the candidate (27 percent) as their highest determining factor, followed by Supreme Court nominees (20 percent), which one may argue covers a bevy of value-centric issues, religious freedom (12 percent), abortion (10 percent), and at the bottom, improving the economy (6 percent), national security (5 percent), and immigration (2 percent).[4] Almost the inverse of what the average voter reported in the same survey.

But these polls, which were not outliers,[5] tell a larger story. A story of goals.

It is difficult to view polls of the type I've referenced without questioning just where the priorities of evangelical Christians are these days. It is even more difficult to deny that it points to a self-interest that dwarfs any of the concerns that evangelical leaders had continually assured the public were at the heart of their support of Trump. Concerns they themselves seemed to think were the most important based on the polling of pastors I cited.

What's clear to me is that the average voter who identifies as an evangelical simply does not prioritize goals in the same way as the pastors and evangelists who purport to lead them.

Based on the constant appeals from those leaders to their flocks that abortion, gay marriage, religious liberty, and other values were safe only with a vote for Donald Trump, it appears they are completely unaware of the disparity, or at least don't feel the need to address it.

The goals of the flock simply aren't lining up with the goals of the shepherds. If a pastor tells his assembled flock every week that cursing in the church is wrong, but cursing continues with reckless abandon in the pews, the pastor would not only be right to bring this up regularly, he'd be downright competent.

So with declining membership, decreasing enthusiasm, and polling that reflects a large disparity between the stated goals of Christianity versus the priorities of the congregants, what is going wrong?

I contend that I've been showing you in the pages of this book exactly what's going wrong. As my dad said when I asked him all those years ago why so many people have a beef with Christians: "Because so many Christians are jerks."

Haven't I made the case that at least among America's most vocal Christian segment, the evangelicals, that is precisely what they have been showing themselves to be?

With evangelicals behaving as jerks, how can anyone expect anything other than the loss of younger generations, the departure of members, and a difficulty attracting new adherents to the faith?

As I mentioned earlier, my worldview was shaped in large part by the Christians I was around. My parents, in particular, instructed me on an inclusive version of faith that was welcoming of diverse experiences without being malleable or flimsy in its doctrinal underpinnings.

I was surrounded by proselytizers of Christianity who were consistent and sincerely believed in the Bible, which was, importantly, displayed in such a way that made the connection between their faiths and their lives obvious.

Christians are the face of faith for nonbelievers, so it makes sense that how they present God will have a direct impact on how someone unfamiliar or unbelieving might see Him. Which is, if you think about it, an incredible responsibility that should be taken very seriously.

The Fixed Point Foundation—an organization that existed for the purpose of, as they explained on their Facebook page, defending the Gospel "in the secular marketplace and equip[ping] others to do the same"[6]—interviewed college atheists back in 2013 to figure out what made them leave the faith.

An article in the *Atlantic* recapped some of what was discovered, but the answers these young atheists gave all seemed to lead to the same conclusion: the proselytizers of Christian faith were the primary influencers.

From the *Atlantic*:

> *Slowly, a composite sketch of American college-aged atheists began to emerge and it would challenge all that we thought we knew about this demographic. Here is what we learned:*

THEY HAD ATTENDED CHURCH

> *Most of our participants had not chosen their worldview from ideologically neutral positions at all, but in reaction to Christianity. Not Islam. Not Buddhism. Christianity.*

THE MISSION AND MESSAGE OF THEIR CHURCHES WAS VAGUE

> *These students heard plenty of messages encouraging "social justice," community involvement, and "being good," but they seldom saw the relationship between that message, Jesus Christ, and the Bible. Listen to Stephanie, a student at Northwestern: "The connection between Jesus and a person's life was not clear." This is an incisive critique. She seems to have intuitively understood that the church does not exist simply to address social ills, but to proclaim the teachings of its founder, Jesus Christ,*

*and their relevance to the world. Since Stephanie did not see that
connection, she saw little incentive to stay. We would hear this
again.*[7]

As one of the interviewees put it, "Christianity is something
that if you *really* believed it, it would change your life and you
would want to change [the lives] of others. I haven't seen too much
of that."

In other words, matters of tone, rhetoric, and consistency from
Christians are not some minor quibble. People, at least those interviewed here, were hungry for the type of connection between
faith and action that I saw in my parents; and that connection, evidently, was missing for these interviewees.

But in keeping with the theme of this book that it is the very
proselytizers of our faith who continue to embrace the world at
the cost of persuading people toward God, the very person who
conducted these interviews, Larry Taunton, stepped down as the
head of Fixed Point in 2018. The reason? Inappropriate extramarital affairs with two women on his ministry staff.[8]

Is there any wonder why people lose faith when stories like this
are so exhaustingly common?

Leadership matters, as these interviewees made clear. But it's
not just within a think tank's experiment that this is proven true.

Persuasion Is More Difficult

While Christianity, and the evangelical movement specifically, has
always had its fair share of critics, that criticism has gotten louder
in recent years, concurrent with the rise of Donald Trump and in
light of the fact that evangelicals were instrumental in his election,

choosing Trump over Hillary Clinton by a margin of 80 percent to 16 percent, respectively.

Whereas it was not unusual to hear concerns about evangelical influence in politics over the last few decades, the sheer magnitude and anger directed at evangelicals during the Trump years seems reflective of a shift. To put it bluntly, people are pissed off at evangelicals.

It seems that for those outside of the evangelical movement, many of whom already had suspicions about the sincerity of evangelical Christians, Trump has become a litmus test.

In January 2018, a viral open letter to evangelicals from liberal pastor and author John Pavlovitz said the following:

> *Your willingness to align yourself with cruelty is a costly marriage. Yes, you've gained a Supreme Court seat, a few months with the Presidency as a mouthpiece, and the cheap high of temporary power—but you've lost a whole lot more.*
>
> *You've lost an audience with millions of wise, decent, good-hearted, faithful people with eyes to see this ugliness.*
>
> *You've lost any moral high ground or spiritual authority with a generation.*
>
> *You've lost any semblance of Christlikeness.*
>
> *You've lost the plot.*
>
> *And most of all you've lost your soul.*[9]

Columnist and author Michael Gerson opined at the *Washington Post* that evangelical leaders "have become active participants in the moral deregulation of our political life. Never mind whatever is true, whatever is honorable, whatever is right, whatever is of good repute. Some evangelicals are busy erasing bright lines and destroying moral landmarks. In the process, they are associating evangelicalism with bigotry, selfishness and deception. They are playing a grubby political game for the highest of stakes: the reputation of their faith."[10]

Evangelical Democrat Randall Balmer lamented that the religious right has "dropped all pretense that theirs was a movement about family values."[11]

Just after the 2016 election, writer Patrick Kampert bemoaned in a column in the *Chicago Tribune,* titled "After Trump I Can't Relate to My Evangelical Faith," that "with their embrace of Donald Trump, white evangelicals have lost all credibility, every last shred of it. Jesus said that the world at large would know his disciples by their love, but I see judgmental attitudes and hate where there should be empathy and compassion. I see little resemblance to the Savior we purport to serve." He went on to cite Romans 12:9, "Love must be sincere. Hate what is evil; cling to what is good." And he added, "This has not been followed."[12]

It's easy to imagine that non-evangelicals looking in on this find it all to be quite in line with their preconceived notions of the movement. Perhaps even of Christianity as a whole.

Michelle Goldberg summarized this in the *New York Times,* saying, "Christian conservatives may believe strongly in their own righteousness. But from the outside, it looks as if their movement was never really about morality at all."[13]

Phil Zuckerman, professor of sociology and secular studies at Pitzer College in Claremont, California, wrote for the *Huffington Post* about evangelicals in early 2018, saying, "We can clearly see Evangelical Christianity for what it really is, at least in its North American, early 21st century incarnation: immoral, uncaring, and blatantly harmful."[14]

After Trump won, Kathryn Freeman, director of public policy for the Christian Life Commission in Austin, Texas, said she was "reeling and still in shock. It has been disorienting to see so many evangelicals—including the ones who say they care about racial justice and esteem women—choose a candidate whose message and language was so demeaning and in some ways downright hateful to those groups."[15]

An important point to consider in light of all of this criticism would be whether there's any basis for it or whether it's simply typical partisan posturing and hyperbole. And further, what makes evangelical support for Donald Trump so polarizing that it creates this type of reaction?

Doesn't Trump's accomplishment of evangelicals' political goals outweigh all of this?

Evangelical leaders question why individuals like myself or Russell Moore or the percentage of evangelicals who couldn't vote for Trump had such a problem with this particular candidate, especially given the fact that we've all grown quite accustomed to seeing politicians of low character win elections.

Speaking personally, it was always more than the fact Trump was brash or likely to hurl unwarranted insults at his adversaries. It was never just that he was prone to exaggeration at times or that he was often found to be flat-out lying in ways that were easy to identify.

Donald Trump is far too easy of a target if this book existed solely for the purpose of identifying how he is unfit as a moral leader. His unwillingness to show any depth of self-reflection or humility speaks for itself, and, as a fellow moral failure who has done and still does the hard work of trying to better himself, I find that celebrated stubbornness of his public persona repellent and unworthy of the presidency.

But besides, the question of his value to conservatives and evangelicals is a perfectly legitimate one in light of some of what he has managed to accomplish.

For instance, opinions, mostly divided along familiar partisan lines, vary on Trump's direct involvement in what remains a mostly strong economy. Democrats argue that the underpinnings are poised to bust and that all the strong indicators are leftover glow from Obama's economic policies. Republicans, meanwhile, contend that rampant deregulation along with lowering taxes are sending the economy to new heights.[16]

On abortion, by reversing an eleventh-hour executive order from President Obama in January 2017, Trump opened up the ability for states to opt out of funding non-abortion-related services at any Planned Parenthood that still performed abortions. Additionally, he appointed pro-life justice Neil Gorsuch to the Supreme Court to fill the vacancy caused by Justice Scalia's passing and, more recently and controversially, Justice Brett Kavanaugh, who took the seat previously occupied by Justice Anthony Kennedy.

Additionally, at least insofar as his supporters are concerned, Trump has been adequately "tough" on terrorism and ISIS.

The primary reason that these accomplishments don't sway me is because I never doubted the possibility. I doubted that it was a certainty, but I would've been a fool if I had not considered during the 2016 election whether or not a President Trump would be capable of getting some things done.

During the election cycle I cited the inability of conservatism (much less evangelicalism) to persuade voters in the future, on that fateful day when power inevitably changes hands. What would the future of conservatism, the GOP, and American evangelicalism look like if we were most recently remembered for wholeheartedly championing a man of untold chasms of depravity as our leader?

On the morning after the election I explained this, saying, "I believed, and still believe, that Donald Trump's character is such that he will drive this country further into problems we may have been able to escape and that the Republican Party will do nothing to stop him, thus destroying the credibility of conservatism as a philosophical force in American politics."[17]

I have issues with Trump himself, but *the* issue with Trump is less about what *he* does and more about what *other people* do on his behalf or in opposition to him.

As I saw it at the time, his existence seemed to create a level of partisan-driven intellectual dishonesty among his adherents (and his detractors) that seemed destined to drive the divisions in America deeper than ever. However, at that time, and till today, my primary concern above all else is how he seems to have this same effect on the proselytizers of my faith. He pulls many evangelicals into a vortex of moral ambiguity and relativism that has become almost required to continue supporting him.

He so often and so loudly offers the opposite of what a Christian leader, or even just a competent leader, should, and as a result of his position it is impossible for those who support him to avoid being asked, What do you think of this thing he has said or done?

What has been so sadly remarkable is that, consistently, rather than rebuking him or even simply offering indifference, he is applauded.

As sad as the distinction may be, it would be one thing if this man was continually finding disappointment among the base for his actions, *even if* they still supported his agenda or accomplishments. Instead, we have moved into a world where the man's worst aspects are reveled in; his every unthinkable action is celebrated and shared.

When faced with overwhelming evidence of a wrongdoing by Trump, evangelicals from the grassroots level all the way up to leadership repeatedly show themselves to believe the unbelievable or excuse the inexcusable. Even in the rare cases where they are forced to acknowledge that something he said or did was objectively wrong, they seem very quick to patch it up on his behalf, despite little to no effort being shown on his part.

To put my overall issue with his candidacy and his presidency more succinctly, Donald Trump is the single greatest source of hypocrisy I've seen in a movement already perceived by many as pharisaic.

The contribution this unyielding defense of the indefensible offers to that unfortunate but common perception of Christianity is deep.

While throughout the election and early in his presidency I was told, "We aren't electing a priest," as a way of deflecting any moral expectations of Trump beyond the implementation of an agreeable agenda, polling seems to suggest Trump is having an impact directly on our very understanding of morality, at least among Republicans.

A Quinnipiac University poll found that 67 percent of Americans do not believe Donald Trump is a good role model for children. But as a subset, those who identify as Republican say he *is* a good role model—by a margin of 72 percent to 22 percent.[18]

This is especially important when you consider that, according to Quinnipiac, "There is almost no gender gap in grading President Trump's standing as a role model. Every party, gender, education, age and racial group, except Republicans, say the president falls short."

When you consider the implications, it's staggering. Standing against the wind of almost universal consensus that Trump lacks the moral standing to be a good role model, Republicans, the party the vast majority of evangelical Christians call home, defiantly stand with him.

Again, the issue I'm addressing is *not* whether Trump is, in fact, a good role model, though I believe he is not. It is that Republican voters, who have long identified as value driven, seem to have become flexible in their understanding of morality, all in the service of defending one man.

One of the earliest ways we knew of this so-called Trump effect came to light in 2016. Five years earlier, only 30 percent of white evangelicals polled believed immoral elected officials could still perform their jobs ethically. As Trump's popularity rose, that number seemed to shift to allow support of a man with his well-known character defects, rising to 72 percent of white evangelicals now believing immoral elected officials *could* perform their jobs ethically.

This was the evangelical community adopting the "none of our business" philosophy that they had so derided when the Monica Lewinsky scandal was unfolding.

It reminded me of President Bush's 2008 comment that the bank bailouts had been his attempt to save the free market by abandoning free market principles. Despite how much I liked Bush, that comment never sat well with me. Either the truth is true or it is not.

Employing that same faulty reasoning, American evangelicals, allegedly the frontline warriors in the culture war, seemed to be saying that the only way to save America from depravity was to follow the depraved.

What I believed of evangelical support at the time of Trump's election was that evangelicals were diminishing their ability to persuade, something the movement was already fighting an uphill battle to do in the first place. Persuasion requires some measure of credibility, which, as is being demonstrated daily in the news and at the polls, evangelicals are sorely lacking.

It wouldn't be especially hard to dismiss many of the criticisms referenced earlier in this chapter as typical of religious bigots or people not particularly empathetic to Christianity in the first place. After all, not all of the people I quoted are themselves evangelicals. But to do so would be monumentally missing the point. The entire basis of persuasion is that a person, even a critic, needs to be convinced. Christians aren't tasked with persuading themselves, they're tasked with persuading others. Yes, many of the criticisms come from outside Christianity. But it is those outside Christianity whom one would think are the most important ones to convince.

Put simply: it makes no sense for Christians, tasked with witnessing to the world, to offer no concern about how the world perceives them.

The Bible makes this abundantly clear in verses such as 1 Peter 2:13–15:

> Be subject for the Lord's sake to every human institution, whether it be to the emperor as supreme, or to governors as sent by him to punish those who do evil and to praise those who do good. For this is the will of God, that by doing good you should put to silence the ignorance of foolish people.

Part of what is implied in this verse is something that is dealt with in politics regularly. You don't give ammo to your enemies. Don't cede the battlefield by living up to the worst expectations of those that oppose you. When you do this, you drive people away from being persuaded.

Of course, it may be reasonable to wonder why I would believe that those who currently hold the majority of power in this country, and managed to maintain much of that power in the recent midterm elections, would need to view their ability to persuade as a concern. But as the last two years or so since that fateful 2016 election night have shown, persuasion is precisely what has been lost.

Our political discourse has become more divisive than at any point in recent history. Under President Bush we saw massive protests and antiwar demonstrations. Under President Obama, both the Tea Party and the Occupy Wall Street movements swept the country.

But at no time in my memory have I witnessed the divisive bitter hatred for one another that I've seen under this president.

And I'm far from the only one who sees it that way. An NBC/ *Wall Street Journal* survey found that eight out of ten people polled believe America is "mainly" or "totally" divided.

A full 90 percent of Americans polled believe that the divisions between the two major political parties are a "serious problem," meaning that despite all the *ways* that Americans are divided, their one point of agreement is that the *level* of divisiveness is a serious problem.

In some cases, I fault the president himself, as well as his supporters and sycophants. In others, it is the people who despise him with a potency so strong it has rehabilitated their views of past objects of scorn like President Bush and Mitt Romney.

I have not seen many members of the so-called Resistance finding a way to persuade people to their cause. In fact, Trump himself remains their biggest recruiter simply by persuading people to oppose him.

I have also not seen many who support Trump rallying people to their views, though amusingly, it is again the opposition that tends to drive people who might otherwise despise him to find defense within themselves in light of oftentimes absurd reactions from the Resistance.

Things are so divided that Americans seem only to be able to find common ground in opposition. Voting against what they hate, not for what they love. This makes for a very cynical electorate and news cycle, where referring to conditions as a "dumpster fire" is just repeating the common wisdom.

People hate each other, and it isn't getting better.

But while the matters of the world should concern us while we are here, it is the ability of evangelicals to persuade people to enter into relationship with Christ that most concerns me. And of late, and even prior to Trump's ascent, that movement has not found much in the realm of new membership.

Moral Welfare

In addition to all of the factors I've already mentioned as to why faith is struggling to keep its members, there is the pervasive issue of moral welfare.

If I were to define moral welfare, I'd say it is the dependency of American Christian churches on government to influence cultural norms in a way that is in line with their Christian values.

To my memory, no single issue in recent history better defined how counterproductive moral welfare can be than the legalization of gay marriage. I'd say there is certainly a biblical foundation for discussion of homosexuality just as there is a biblical foundation for discussion of premarital sex, or any number of other decisions and lifestyles of which the Bible disapproves, or does not appear to approve. But unlike the issue of gay marriage, evangelicals were not so often seen to be marching against the scourge of unmarried cohabitation or looking to pass laws to prevent the activity, despite the fact that they likely found it as much of a sin as they viewed homosexuality to be.

If asked about their opinion of it, certainly evangelicals might easily have answered that they thought premarital sex was a "bad idea" or a contributing factor in cultural decline. But they likely knew several people who were unmarried yet living with their partner and, I'd wager, had even engaged in premarital sex at some point in their own lives.

But of course, in light of this comparison, which I often made at the time, gay marriage was at the forefront for them because there were specific "actions" being taken to codify the issue. Actions the evangelical movement deemed, by and large, to be a taxpayer-funded endorsement and an approval of a behavior they knew to be sinful.

Even granting that this concern was the primary motivation, I had the recurring thought (though the full realization of those recurring thoughts remained elusive until the era of Trump) that the primary inconsistency was exactly what I saw in Jerry Falwell Sr.'s response to the September 11 attacks all those years ago, wherein his immediate reaction was to blame "the gays and the lesbians."

His response lacked any humanity for people who differed from him or led lives he found incompatible with his values. While he seemed able to find patient, compassionate understanding for those whose situations he could relate to from personal experience, there was an obvious absence of empathetic reserves for anyone else.

I attempted to convince evangelicals that perhaps the issue of the "implicit endorsement of sin" through marriage licenses of gay couples was really more an example of the dangers of government meddling in an institution we considered religious. That focusing on removing the opportunity for such a conflict to occur was a better path forward. Perhaps one that even those opposed to the evangelical view of gay marriage might embrace.

I often summarized by saying, "Perhaps instead of asking the government to give or withhold the keys to marriage we should stop granting them the authority of key master."

This was most normally met with disagreement on the grounds that it was our Christian duty to ensure that government upheld our Christian values, even to the exclusion of others.

It seemed logical to me as a Christian that "behavior modification" is not how you win hearts and minds, much less bring anyone to Christ. To bring individuals to Christ, you discuss with them God's deep desire to have them be in a personal relationship with Him. Whatever might be placed on their hearts as an outward expression of that relationship is something that the persuasive evangelical can merely observe. Perhaps that's why we call it "witnessing."

But instead, evangelicals fought against gay marriage so fiercely almost entirely through governmental solutions that it was hard for anyone on the other side of the debate to hear anything other than homophobia.

By 2015, gay marriage was legalized at the Supreme Court in every state, and the popular cultural view of evangelical values was more diminished than ever, with many seeing evangelicals as relics of the past. Fair or not, "clinging to their gun and their Bible" had gone from offensive gaffe to common wisdom.

As an evangelical myself, I thought it would've been far preferable to lose such a public battle in the culture war on the merits (or at least in spite of the merits) of the argument. But the merits of the argument had been beaten to death, not by proponents of gay marriage, but by evangelicals themselves.

They didn't lose the argument. They never even made it.

They didn't offer a compelling vision of the institution of marriage, nor did they spend any significant time cultivating a reasonable conversation with adversaries. Instead, they went straight to the government and demanded action while offering condemnation absent any sense of love or understanding from the pulpit. Pulpits that, I will remind you, were all too often used by leaders who secretly lived scandalous secret lives or by other leaders who stood by and watched it happen without offering the same kind of condemnation they reserved for the "heathens."

They were so committed to the notion that they were at war *with* the culture, not fighting for the betterment *of* the culture, that they ultimately offered the concept of Christ's compassion as little more than an asterisk. A side note. Their first objective, as far as those whose ears they'd hoped to bend were concerned, was crushing anything they didn't understand with no regard for beliefs that might exist outside their own worldview.

Jesus demonstrated over and over throughout the Gospels the importance of compassion and understanding. Whether in Luke, chapter 15, where he gathered with the tax collectors and "sinners." Or even of His own executioners, whose cruelty prompted Him to beseech the Father, "Forgive them, for they do not know what they are doing" in Luke, chapter 23.

These verses are examples of how we should all approach sin and sinners: with the compassion, through words as well as actions, that God showed us first.

And apologies if anyone finds the idea offensive, but I do not picture Jesus taking the stage with a megaphone to declare cultural jihad on homosexuals. The comparison of such a visual to the destruction of Sodom and Gomorrah is but one in a long line of incomparable analogies that I've heard evangelicals use in defense of such a bold view of Jesus's politics.

They all revealed the same thing to me, which has only become more prevalent when faced with the moral conundrum of wanting to elect someone you know Jesus probably wouldn't campaign for: evangelicals no longer saw themselves as witnesses. They saw themselves as righteous.

And armed with their self-righteousness and certainty, they sought to grow the kingdom, not through emulating Christ but through moral welfare.

The founders certainly saw and spoke a lot about their belief that government might reflect the religious values that they held as Christians. But even in reviewing their thoughts, I've not seen any evidence that they hoped religious institutions would abdicate their responsibilities in favor of government force.

It may be strange to hear a conservative evangelical Christian playing footsie with what sounds like a "separation of church and state" argument, but I think the truth of what the founders intended is important in the context of the "loss of the culture" that is often decried by conservatives.

In 1798, in an address to the Massachusetts militia, President John Adams said, "We have no government armed in power capable of contending in human passions unbridled by morality and religion. Our constitution was made only for a moral and religious people. It is wholly inadequate for the government of any other."

This quote might be used as an example of how there was an expectation that the government was responsible to uphold Christian values.

He also said, "The experiment is made, and has completely succeeded: it can no longer be called in question, whether authority in magistrates, and obedience of citizens, can be grounded on reason, morality, and the Christian religion, without the monkery of priests, or the knavery of politicians," in 1788.

But before you assume that he is saying something akin to "Yes, please legislate Christian values!," consider these quotes by the same man.

Nothing is more dreaded than the National Government meddling with Religion.

—JOHN ADAMS IN A LETTER TO BENJAMIN RUSH IN 1812

Although the detail of the formation of the American governments is at present little known or regarded either in Europe or in America, it may hereafter become an object of curiosity. It will never be pretended that any persons employed in that service had interviews with the gods, or were in any degree under the influence of Heaven, more than those at work upon ships or houses, or laboring in merchandise or agriculture; it will forever be acknowledged that these governments were contrived merely by the use of reason and the senses.

—JOHN ADAMS, WRITING IN HIS THREE-VOLUME POLEMIC TITLED *A DEFENCE OF THE CONSTITUTIONS OF GOVERNMENT OF THE UNITED STATES OF AMERICA*

These two additional quotes shed a different light on the previous, without undercutting or contradicting them. And while four quotes said over two decades certainly do not define the entire works or philosophies of a man, it is not difficult to find a thread that speaks to something far different than the legislation of Christian values.

That common thread is the people. Adams seemed to believe, as a Christian, that a government composed of individuals who seek to uphold Christian values is served best through the American Constitution. Furthermore, that such a government is the only one that *could* adequately uphold the values they'd set forth in the Constitution, which seems reasonable since the framers' own Christian values instructed its writing.

But all of this resides within a structure that is dependent upon government being representative of the people they govern. In other words, government would reflect Christian values *if* the people to which the government answers are themselves a Christian people. This is a moral view, and a political view, but it is also, obviously, a supremely logical one.

To expect that flow of philosophical influence to work in the reverse, with the government elected to mandate Christian values as law, would not serve to foster a Christian culture. It would incite rebellion and cultural decline. And I'd argue that in many ways it has.

Naturally we are not in these pages going to debate the finer points or historical perspectives of each of the Founding Fathers at length, nor do I represent here that this small sample of words from a man's entire life can be considered to represent him as a whole or his thinking in total. They cannot do that any more than my past words or actions can be considered the summation of my own self or identity.

Whether the founders endorsed or rejected the idea that the cultural influence of the electorate will influence how the government conducts itself, the fact is, it will. And the Bible instructs us to do exactly that.

One could be forgiven for having confusion about God's view of the intersection of government and religion. In several places, the Bible very specifically directs believers to simply submit to the authority of man.

ROMANS 13:1–2

[1] Let everyone be subject to the governing authorities, for there is no authority except that which God has established. The authorities that exist have been established by God. [2] Consequently, whoever rebels against the authority is rebelling against what

God has instituted, and those who do so will bring judgment on themselves.

In these verses, Paul specifies that no government authority is in its position without God's appointment. It is important to understand that appointment and endorsement are not the same. Entirely in line with previous verses I've quoted, this is the manifestation of God's control over ends. These authorities exist to fulfill *God's* design, not ours. And thus, even under the rule of an oppressive government such as the Roman Empire, God's command is that we first obey His expectations of our character.

We are called to be the subjects of God primarily, which means endeavoring to follow His righteous commands ahead of whatever reward may lie at the end of disobeying Him through rebellion.

This might raise questions, such as whether or not the American Revolution was disobedience to God's commands. Is God suggesting we live under the rule of tyranny without seeking to usurp corruption? All in the name of upholding our own consistency of character? Not quite.

The Bible goes to great lengths to make clear that while a Christian is to submit to earthly authority, it is obedience to God that is the motivation, not devotion to those leaders. And if those authorities require you to defy God's ultimate authority, you are to consistently place your faith and allegiance in Him, even in defiance of that government.

This is illustrated in the book of Daniel, chapter 6, when, as a high-ranking member of government under the Persian king, Darius, Daniel was ordered to sign a document that would punish anyone who did not worship the king, specifically that "whoever petitions any god or man for thirty days" should be executed.

This would have directly usurped God's authority and sovereignty, and so Daniel flatly refused, earning him a place in the lions' den. And as every child of Sunday school already knows, Daniel was spared execution through God's intervention.

So how do we reconcile what appears to be God's express command to view obedience to government as obedience to Him, while simultaneously instructing us to disobey government in certain situations? How are we to ascertain the differences and be sure of what actions would constitute obedience?

Honestly, it only sounds complicated. Placed in the context of human history, it's actually quite simple.

There is a reason that "I was just obeying orders" has been used as a way to hark back to Nazi murder. Obedience to chain of command is not only a vital and lifesaving aspect of military conduct, it's also commonly considered honorable and morally right. This is accepted without question by most conventional standards.

But "I was just obeying orders" illustrates another uncontested truth: there are things that are of greater importance and should supersede even that which is otherwise lawful. In the case of Nazis' committing atrocities in the name of obedience, their higher devotion should have been to humanity.

Even smaller examples still highlight this entirely consistent philosophy.

As the owner of a video production company, I've made hundreds of videos for clients. In some cases, I disagreed with what my client wanted to do or thought it was not presenting their message in a way that would communicate what they'd hoped.

Though I'd voice my concerns, it was my job to do what I was being paid to do. And so, I'd relent on a dumb call in submission to the will of my client. The old "counsel, then advocate" protocol I've heard lawyers use, which indicates that you make your case, but ultimately you will service the needs of the client.

However, while it has yet to happen, were a client to bring me a video script that I believed defied my own principles, those same principles that instructed my "counsel, then advocate" mentality previously would then cause me to refuse the project and risk severing my relationship with the client.

I can proudly say that I will dutifully fulfill the needs of my client while simultaneously having the capacity to refuse their requests when it is in defiance of my higher obligations.

In my example, have I contradicted myself? Have I abandoned my underlying principles? Of course not. Neither did those who believed in military command structures that still prosecuted Nazi subordinates. And neither did Daniel.

God has certainly shown and commanded that obedience and rebellion can both be acts of submitting to God's authority, just as they can both be acts of sin.

But the story of Daniel has another lesson we should heed that furthers my point that government should not attempt to force a cultural outcome through law.

Following Daniel's miracle in the lions' den, King Darius celebrated his survival. But, more important, he witnessed firsthand Daniel's faith and God's protection of him, and it changed Darius.

Through Daniel's observable faith, not words or decrees or promises, King Darius declared that the God of the Bible is the "living God" and he submitted to "His dominion" over all and for eternity.

The government did not change Daniel, though it tried. Daniel changed the government. Not through law. But through enduring faith.

American evangelicals attempted, and still attempt, to control a multitude of behaviors that contribute to the overall cultural and moral decline of America, by appealing to authorities of man instead of the sovereignty of God.

Either Christians learn to accept this lesson or they offer compelling reasons for its fault. But, excluding the latter, their devotion to the principle of an inerrant Word of God should dictate how they approach every issue where the involvement of religion is working against God's wishes under the guise of obedience.

Moral welfare is a corruption of God's will as it relates to our interaction and influence over the culture. It is an abdication of responsibility and a perversion of our purpose.

At this point it's worth looking back and acknowledging as the author that a lot of what I've laid out has been the problem, not the solution.

I've established that evangelicals, the loudest and most politically engaged portion of American Christianity, have vocally and in view of everyone publicly abandoned their stated principles, principles I also established their many critics doubted were earnestly held in the first place.

I've shown how their devotion to this path has become a religion of its own, the New Good News.

I've shown how starkly it contrasts with their insistence of character demanded in the 1990s when Bill Clinton was the country's immoral symbol, and I've demonstrated how the culture war has been used as both a source of persecution complex and a mask to project noble ends onto what amounts to a pursuit of worldly acquisition.

I've further cited the statistics that show that not only are evangelicals declining in membership, but that faith on the whole is declining. That persuasion is becoming more difficult as Christians are increasingly seen as aligning with the caricature people have always had of them.

And I've revealed how far apart the goals of congregants are from the leaders that preach the gospel to them, not only in that

196 | The Immoral Majority

their priorities are far less about religious freedom and morality and more about tax cuts and deregulation, but also in that their understanding of morality has shifted alongside the rise of Trump evangelicalism and the New Good News.

So what are Christians to do?

How do they navigate the circumstances of an election without falling into a "lesser evil" dilemma? How do they go about maintaining their reasonable worldly wants and needs without subsequently forgoing their eternal obligations?

It may seem strange that I have reserved only one chapter to answer so many complex questions, but that's because the answer is and has always been simple.

Not easy perhaps, but simple.

Trust. God.

Chapter 7
TRUE VICTORY

After everything you've read so far it may surprise you to know that I don't really care if you support or voted for Donald Trump. I'm also not here to convince you not to vote for him or support him now.

If you wake up every morning and put on your Make America Great Again hat, watch the president tussle with the press, and thank God that someone is finally standing up to the elites, by God, you be you.

I'm also not here to contend that to do all those things is inherently un-Christian or blasphemous or that Trump's presidency is the manifestation of Satan's dominion over our nation.

If God's kingdom were dependent upon the actions, words, and pathologies of Donald J. Trump, then, frankly, we have a pretty weak God, and I'm kind of counting on the idea that God is in control.

As I've made clear, I couldn't and didn't vote for the man. And it's true that I spent significant time and effort trying to prevent his nomination and outwardly calling for his defeat in November, despite my distaste for Hillary Clinton and despite my belief that a Donald Trump presidency would far better serve my short-term policy wishes than a Hillary Clinton presidency ever could.

In fact, I predicted that I would like a great number of his policies if he were to win, and this has proven to be true two years into his term.

As a conservative, I was pleased with the appointment of Justices Neil Gorsuch and Brett Kavanaugh to the Supreme Court.

As of this writing, unemployment has reached record lows, the economy is booming in a lot of ways, his approval ratings across normally Democratic voting blocs have risen almost as quickly as Americans' standard of living and our economy's gross domestic product.

And these are far from the only aspects of the president's policy moves that I've been pleased to see implemented, nor will they be the last. These are all good things, and I won't pretend they aren't.

But does the good of a President Trump outweigh the bad of a President Trump?

That all depends on how one measures good and bad, doesn't it? Certainly, his election and presidency have been good for the things I've named, but the costs remain.

The issue is not whether a professed Christian supports Trump. The issue is about the reasoning employed to get to that point. What did they put aside, rationalize, or deny to themselves in order to vote for him?

If the answer to that question is "absolutely nothing," then there's not much for me to say about that. We disagree, but I would take you at your word that you seriously considered the same things I considered and simply came to a different conclusion. But regardless of the number of people who would immediately tell me that their Christian conscience is and has always been clear, it is painfully obvious that this is not the case for everyone.

The level of rationalizations and relativism that is constantly being preached and expressed in favor of this president's support is strong evidence that there are plenty of people who struggle with whether or not they have been honest with themselves and, conversely, with God.

As I quoted in chapter 1, the Bible says, "For what does it profit a man to gain the whole world and forfeit his soul?"

What I have had to ask myself—beginning in 2016 and continuing until today—is whether the profits of this presidency would outweigh the costs to my soul that would come with excusing or ignoring or rationalizing things that I felt certain I could not support.

Did God at any point in the entirety of the Bible ask His followers to choose between immoralities? Where does practicality end and religious conviction begin? This is the burning question that has remained with me throughout Trump's presidency. Am I being impractical in my ongoing opposition? Why would I turn down the opportunity to see the fruits of conservative policy even if they were implemented by a man I personally find detestable?

But of "lesser evils," the first reasonable question to ask is, does such a moral calculus even exist?

Turns out, it does.

Moral Hierarchicalism

I first saw the Disney film *Aladdin* at about age fourteen. An age when I had already mostly outgrown cartoons, but, living in a household with two parents who had helped open Disney World in the 1970s and a sister who aspired to be a Disney cartoonist, I suffered through the film and ended up enjoying it despite myself.

The opening scenes of the film include a musical number in which the titular character is busted for stealing bread. As the authorities chase him through buildings and alleyways, Aladdin cheerfully sings the line, "Gotta eat to live, gotta steal to eat!"

It really is a reasonable concept.

Stealing is wrong. The Bible says it. Man says it. Kids' shows say it. Even country music mostly agrees.

Then again ... survival is morally right. And certainly in this instance, assuming that Aladdin is genuinely starving, stealing may be his only option. Stealing is the "lesser evil," he sings.

Is he right?

This conflict of wrongs, if you will, is an easy scenario to concoct, and the dilemma does occur in real life: a situation where it's less about creating false limited options and ignoring other possibilities, and more about being stuck with those options.

If someone were on the top floor of a burning building and there was little chance they could survive jumping out of a window to escape the flames, yet the fire had become so close that it was on the verge of engulfing them, would jumping out of a window to spare themselves death by fire really be anything close to committing the sin of suicide? In this scenario, aren't they really just choosing which would be the less painful version of imminent death?

There's no question terrible options exist. People have been faced with these awful choices, and in both of the examples, the answer is that their choice was *not* sinful.

Jumping out of a window with no chance of survival in order to escape engulfing flames or choosing to steal food as a way to prevent dying of starvation are not examples of immoral choices, nor are they examples of false choices.

In fact, these conundrums don't have to be nearly so dire to be a valid example of someone choosing between what could be described as "lesser immoralities" or, more commonly, "lesser evils." But at least some level of urgency should exist, and that urgency cannot simply be fear.

There is an ethical hierarchy that Christians should and do employ in these situations.

Renowned apologetics pioneer and Christian ethicist Dr. Norman Geisler has written extensively about the concept, but for the sake of brevity I will provide an excellent summary of Geisler's view on hierarchicalism as written by author Dennis McCallum.

"In [ethical hierarchicalism], one ought to do whatever fulfills the highest moral rule in a situation. When this is the case, such action is right, and the person in no way does wrong. Under this view, there are no tragic moral dilemmas. The lesser of two evils is a misnomer, it is argued, because such lesser evil is actually good."[1]

Basically, the principle here is that when you are on the brink of starvation and have the opportunity to save yourself, the superior moral good of staying alive supersedes the lesser immoral act of stealing. In this sense, the "sin" committed in favor of a greater moral outcome is actually a moral good in and of itself.

Seems a bit confusing and ripe for abuse, doesn't it?

For anyone who does not subscribe to the idea that there is a God who is interested and involved in His creation and whose morality is absolute, relativism probably makes a lot of sense. The ethics of a situation can be based purely on the situation at hand and a preferred outcome. There is no divine consideration to draw into your reasoning or that might compel you to act counter to your desire in favor of a larger purpose.

For Christians, it's not so easy. We are tasked with accepting God's perfect moral standard no matter the cost to ourselves if necessary.

In *Christian Ethics*, Dr. Norman Geisler lays out why the moral hierarchical structure, called graded absolutism, is a theologically sound method through which Christians can determine whether they are truly facing a dilemma in which they may have to choose a lesser of two evils.

Geisler writes, "In real, unavoidable moral conflicts, God does not hold a person guilty for not keeping a lower moral law so long as one keeps the higher law. God exempts one from his duty to keep the lower law since he could not keep it without breaking a higher law."

He explains:

> The graded absolutist does not proclaim that the evil is a good thing to do, but rather that the highest obligation in the conflict is the good thing to do. For example, in falsifying to save a life, it is not the falsehood that is good (a lie as such is always wrong), but it is the act of mercy to save a life that is good—despite the fact that intentional falsification was necessary to accomplish this good.
>
> In other words, it is unfortunately true that what is called "evil" sometimes accompanies the performance of good acts. In these cases, God does not consider a person culpable for the concomitant regrettable act in view of the performance of the greater good.
>
> In this respect, graded absolutism is similar to the principle of double effect, which states that when two results—a good result and an evil result—emerge from one act, the individual is held responsible only for the intended good result and not the evil result that necessarily resulted from the good intention. For example, a doctor who amputates to save a life is not morally culpable for maiming but is morally praised for saving a life.

Geisler concludes on the topic with this summary: "Moral laws sometimes come into unavoidable moral conflict. In such conflicts we are obligated to follow the higher moral law."

Biblical examples include James 2:25, in which Rahab was justified in lying when she hid the messengers and sent them safely down a different path and away from the people who would have killed them. The higher moral law was fulfilled in saving their lives and thus the lower immorality, lying, is exempt.

When Samson destroyed the temple, he was following God's command, even though he was simultaneously technically committing suicide. It would be difficult to conceive of a higher moral law that obeying God's command, so whatever Samson had to do in the service of that obedience is once again exempt. The same is true of Abraham, had he been forced to go through with sacrificing his son Isaac. In Mark, the Pharisees criticize Jesus for healing a man on the Sabbath, thereby doing "work." Jesus clarifies that while the Pharisees are following the letter of the law, they are not following the spirit of it, for the purpose of the Sabbath was not to prevent human flourishing but to encourage it. "The Sabbath was made for man, not man for the Sabbath," He said elsewhere. "So the Son of Man is Lord of the Sabbath."

Thus, there is a higher law to be considered. And that higher law *is* Jesus.

So with this in mind, in order to make that case that voting for Trump is, for a Christian, the fulfillment of a "lesser evil" conundrum, and represents a choice of this higher law, there are several things that have to be established.

First, is an immorality being committed by voting for or supporting Donald Trump? Is supporting a person of demonstrated low character who is antithetical to Christian expectations of leadership sinful? But, most important: Is the focus on the Higher Lawgiver?

God Is Not on Your Side. You Must Be on His.

Despite my belief that Hillary Clinton did not represent someone whom I could truthfully say represented my values, I was ready to accept a Clinton victory when balanced against my fear that

Donald Trump as president would destroy conservatism's ability to persuade.

I reasoned that if he won and became the standard-bearer for what purported to be conservatism, any short-term policy gains from his presidency would be undone by a public who would associate it all with his person. I still believe this is true, but that was not a good reason to vote against my values, which is why I would not vote for her, opting instead to pull the lever for an independent candidate.

When I voiced that I was doing this, and even since, many conservatives told me that I had "thrown away my vote," at best, or, at worst, that I had tried to "get Hillary Clinton elected."

I had done neither thing. I had not only voted my conscience, I had done so in an effort to obey God in the way I understood Him. And I prayed that I had chosen well.

But for many of my detractors, the choice was obvious. There were only two reasonable possibilities and only one that God would be okay with, according to them. If you'll recall, Trump even said so himself! As he told Christian leaders, "Pray for everyone. But what you really have to do is you have to pray to get everybody out to vote for one specific person."

Many people have quoted the verse, "If God is for us, who can be against us?" from Romans, Chapter 8. It's a powerful idea with a meaning that is important for Christians to accept.

If we are to do that which God asks of us, then certainly we are on God's side. And if we are on God's side, then what chance does anyone have in opposing us? None. Whether that plays out immediately or in eternity, God is and always will be victorious, so certainly being on "his side" is objectively good.

But this has shifted into something else. Something more like, "If I'm doing something to gain God's favor, how could He be against me?" This bastardization of the original intent is something I see often.

Christians, in an attempt to "assist" God in achieving His ends, will work feverishly toward a goal, ignoring how often they separate themselves from God while in pursuit. But in their mind, their compass continually points toward this "good end" they believe is in store. An end God would certainly approve of. And as such, "Who could be against me?"

But what trying to get God on your side, as opposed to making sure you are on His side, misses is that God is not on our "side," even when He helps us achieve victory (assuming, of course, that we even recognize victory when it occurs). He is achieving victory in accordance with His design. Not because you have somehow constructed the circumstances that will "allow" it.

There simply is no pulling of a lever in a voting booth that will deny God His purpose when He pursues it, nor is there any pulling of the lever that will earn His allegiance to your "side."

Without this first basic aspect of understanding God's intent as it relates to His people and how they conduct their lives, nothing else in this book will resonate.

God's purposes are higher than ours. This is shown throughout the Bible but nowhere is it clearer than in Joshua, Chapter 5.

13 Now when Joshua was near Jericho, he looked up and saw a man standing in front of him with a drawn sword in his hand. Joshua went up to him and asked, "Are you for us or for our enemies?"

14 "Neither," he replied, "but as commander of the army of the Lord I have now come." Then Joshua fell facedown to the ground in reverence, and asked him, "What message does my Lord have for His servant?"

15 The commander of the Lord's army replied, "Take off your sandals, for the place where you are standing is holy." And Joshua did so.

What is so important to get about these verses is that this commander, a servant of God, even as he is coming to Joshua's aid, specifies that he is not for Joshua any more than he is against him and for his enemies.

In response to that question, he could've simply said yes as a matter of purpose. Certainly he was there to help Joshua. So, in that sense, he is on Joshua's "side."

And yet, he responds with "neither."

Then, as if to recalibrate Joshua's understanding of silly notions like "sides," the commander tells him to take off his shoes in reverence of the holy place he stood. In this, the commander, who is set to assist Joshua against his enemies, has not only made clear that he was not on anyone's side other than God's, but that God's higher purposes are of primary importance, as is the expectation that Joshua revere them.

A series by theologian J. Hampton Keathley III, called "Studies in the Life of Joshua," explains why this moment, as well as the full scope of the story at hand, matters:

> Chapter 5 describes the consecration of the people of Israel in preparation for the great task that lay before them. As such, it stands as a bridge between the crossing of the Jordan and the beginning of the military campaigns to subjugate the inhabitants of the land. For many, however, especially to those trained in military tactics, this chapter may seem like an enigma, at least from man's point of view. And of course, that's precisely the issue here. God's ways are infinitely higher than ours. From all appearances, now was the time to attack the enemy. The people of Israel were filled with the excitement and motivation of having miraculously crossed the Jordan. They apparently knew the enemy was in disarray from the standpoint of their morale (5:1); so surely, it was time to strike.

Many of the military leaders under Joshua's command may have been thinking, "For goodness sake, let's not wait! Let's go! Now is the logical time and the enemy is ripe for the taking!"

But in God's economy and plan there are spiritual values, priorities, and principles that are far more vital and fundamental to victory or our capacity to attack and demolish the fortresses that the world has raised up against the knowledge and plan of God (2 Cor. 10:4–5).

When you consider the implications and how they relate to Donald Trump's election and presidency, it is sobering.

Throughout his rise, evangelicals in varying degrees agonized or at least fretted over the decision. There were many, some of whom spoke publicly, who were unsure of what to do.

How could they vote for this man when he seemed so clearly to defy what they felt certain they should expect in the form of a leader they can trust.

They stressed and questioned themselves over this, finally landing on the side of working within the fallen world to advance God's kingdom with the tools available to them. Ignoring, perhaps, the various ways they might even be defying God, since there were greater considerations at hand and "unchangeable" pragmatic realities.

This was the hand we were dealt, they'd claim. And so, they endeavored to do the best they could, which often involved rationalizing their own decision and offering ways to clear the conscience of others who still struggled.

But it was all premised on this false notion that Joshua's story upends.

By all accounts, Hillary Clinton would, as president, have nominated a liberal Supreme Court justice who would likely have supported the continued legal status of abortion. She would not have taken seriously any claims that religious liberty is being violated in any meaningful way. She would have implemented a great many things that evangelicals have a problem with, either from a Christian perspective or from a perspective mostly rooted in their conservatism.

Donald Trump promised that he would do the opposite. He would appoint pro-life judges; put an end to violations of religious liberty; and do things more in line with conservative ideals, like implement strong immigration rules and tougher foreign policy posturing.

These were the two sides that evangelicals saw. And when confronted with the decision in the voting booth, they said to God, "Are you for Trump? Or for Clinton?"

And as God has shown us through His word, His answer would have been the same as mine and many others who refused this dichotomy: "Neither."

This answer should have left any Christian with a much simpler choice.

If you believed either Hillary Clinton or Donald Trump was someone who reflected your Christian values in their character as well as their promises, then voting for either one was perfectly fine.

But if you had to equivocate and make excuses? Rationalize and deflect? Close your eyes, hold your nose, and pull the lever? And if you did this while telling yourself it is because sometimes you have to defy what God expects of you in order to help Him achieve His own ends?

Then you have erred in your reasoning.

God would have accomplished (and will accomplish) His ends regardless of whether Hillary Clinton or Donald Trump is president. His victory is not our victory. His victory is all Christians voting in accordance with their conscience and in obedience to His values.

You Are Not Guaranteed an Easy Button

The common theme here among Christians and their leaders is their lack of faith in the goals of God, favoring instead their personal goals.

Some may surely believe they are approaching this with the intent of fulfilling God's purposes. But they are clearly doing it in a way that they must on some level be aware God would caution against.

Self-interest, a term I've used several times in this book, is often believed to be a flexible term. Innocuous even, barring any context.

The context can be good, such as "I have a self-interest in eating because if I don't eat I'll die," or "I have a self-interest in career advancement so I can provide for my family."

Or it can be bad, such as "He cheated on his handicapped wife in service of his own self-interest," or "The self-interest of the company's CEO had him embezzling millions of dollars, which ultimately bankrupted the company."

It's often used in conservative discussions to describe profit motive or self-regulation of an industry: "The company chose to lower their prices in light of their competition." This rationale could be described as doing something in the interest of the company, and in a way that's comparable to self-interest, while also not being inherently a bad thing.

So it's understandable why someone—especially a conservative—reading this book may have seen the way I employed the term *self-interest* throughout this book and had a nagging feeling or thought, something along the lines of, "What's wrong with voting in your own self-interest?"

And certainly, the answer is, "Nothing, necessarily."

But while Americans may use the term within different contexts, I found the dictionary definition somewhat fascinating in this regard.

From Oxford's online dictionary, "One's personal interest or advantage, *especially when pursued without regard for others*" (emphasis mine).

While I intend to continue using self-interest flexibly in the ways I described above, I would like to offer the definition I'm using in this chapter for the sake of clarity: "One's personal interest or advantage, especially when pursued *without regard for God*." For me, this is the primary and fundamental issue. And it's one that's hard to grapple with.

When you see self-interests in front of you and you know they are attainable, you might easily be able to rationalize why it would be silly to worry about whether God would give you some side-eye for doing it.

But if this is your thinking right now, and if I haven't sold you on the idea that evangelicals placed a premium on their self-interests that showed a disregard for God, let me make one final appeal. There is not, nor has there ever been, a reason Christians needed to choose between voting without regard for God versus voting against their own self-interest. It is entirely possible to achieve both simultaneously, but you have to decide to do something other than simply push the easy button.

What it really boils down to is a question you must ask yourself: How "bought in" am I to the idea of trusting God?

My father presented me with an analogy several years ago that I will possibly mangle a little bit here, but I think it's a great illustration of the concept.

A man is standing on one side of Niagara Falls when he notices a tightrope extended all the way across the water and attached to the cliff on the other side.

He then sees an acrobat who has placed a large barrel sideways on the tightrope, and she's using it to roll across the rope by running backward on top of the barrel.

It's a death-defying stunt, but she manages effortlessly to traverse the waterfall, landing near where the man is standing.

After dismounting, he claps and remarks, "That was incredible!"

The acrobat looks at him and asks, "Do you think I could jump on this barrel and run backward to roll it across that tightrope all the way to the other side of the falls?"

Confused, the man replies, "Well, of course. I just saw you do it, so I'm sure you can."

"Fantastic," she replies. "Get in."

Now, if you were this man, I feel certain that, like me, you would be hard-pressed to get into that barrel, regardless of your faith in the acrobat's skills.

Why risk it? would be one of your first questions. But even if I added elements to the story to make it imperative that the man cross that waterfall, I'd imagine most people would still have trouble extending enough trust to the acrobat to crawl inside and put their lives in her hands.

It is, I would say, one of the more difficult things to do in life. To trust God with your future. But it is far more frightening when an alternative is staring you right in the face that not only looks more enticing but really doesn't even seem that objectionable.

But as God makes clear throughout the Bible, choosing to do something without regard for Him has far greater consequences than simply making one choice or another. God's warnings are not littered throughout the Bible as a way of giving us a bunch of neat guidelines.

Oftentimes they are warnings, warnings specifically directed at each of us—not that God will punish us, but that He is hoping to help us. That He wants us to avoid the consequences that He alone is capable of foreseeing.

The easy button is an attractive mentality. When confronted with achievable ends with little downside save disobedience to God, choosing outside of the limited scope of your immediate self-interest can be difficult. It's a calculus we all face multiple times in our lives.

It's not simply that we've identified or recognized a potential solution. It's that the solution is far more appealing than simply appeasing God at that particular moment.

God is not silent on the matter, He speaks it to us through His Word.

MATTHEW 7:13–14:

13 Enter through the narrow gate. For wide is the gate and broad is the road that leads to destruction, and many enter through it. 14 But small is the gate and narrow the road that leads to life, and only a few find it.

To put it in a more popular context, "The path of the righteous man is beset on all sides by the iniquities of the selfish and the tyranny of evil men."

It's hard. Not easy. And sometimes it's not obvious. In the matters at hand, it is also unpopular among one's peers. But God didn't promise easy, did He?

This is where we come again to the issue of trust.

Trust is one of the hardest parts of being a Christian. Trust sometimes means going against your own instincts, defying what logic might even tell you, and accepting that whoever or whatever you've placed your trust in will not let you down.

Another parable, one Christians often use to describe what trust in God looks like—or perhaps more accurately, what failure to trust God looks like.

There are many variations on this tale, but basically it goes like this:

A man involved in a shipwreck out on the ocean finds himself adrift, floating in the sea and clinging to a piece of wood or wreckage for dear life. His fate uncertain, he closes his eyes and prays that God will save him.

There on his shoddy raft, the man visualizes all of the glorious ways God might intervene. He could send a flock of birds to miraculously lift his tired body out of the water and carry him to the nearest shore. Perhaps God would send a golden chariot pulled across the sky by winged horses. He did not know what miracle God would bestow upon him, but he knew that God had promised that one needs only the faith to move a mountain for God to do it for them. And this man had such faith.

After spending some amount of time adrift, at long last a small boat pulls alongside him. They toss a life preserver out to him, urging him to grab on to it so they could pull him to safety. But the man's faith would not be deterred. God had a miracle in store for him, and this small piece of manmade inflatable rubber was no miracle. As tempting as it was to grab the life preserver, the man was sure it was a test of his faith in God's miracles.

He declined, saying loudly, "Fear not! God will come through! You will see!"

Eventually, the boat moves on as there were other shipwreck survivors to rescue. As they left, a tiny tinge of fear came over the floating but faithful man, but he pushed it down and reminded himself that God was proud of his unwavering faith and the miracle in store was worth the risk. The rescue boat returned multiple times, but each time the man dutifully declined, certain that God would reward his faithfulness.

As the hours passed, the man continued to pray.

Suddenly, a light shone down on him from above and the wind began whipping the waves around him. Had his miracle arrived? After a moment, it became clear that this was not God's work. The light was from a rescue helicopter. A ladder, tossed out from the cabin of the helicopter, was dropped right next to God's faithful servant.

The man smiled, looked skyward, and, as loudly as he could, yelled toward the well-meaning rescuers, "I wait for God's intervention! Go rescue the less faithful if you can! God will come through!"

But everyone else from the shipwreck had already been rescued either by boat or helicopter. This faithful man was the only one who remained, they explained to him.

This information seemed to bring renewed excitement to the floating man of God. Again, smiling, more certain than ever, he assured them that what God had in store must be glorious. While others had chosen to trust in man, he had chosen to trust in God, and when others saw his reward, it would bring them faith as well!

How honored, he felt, to be used as a miracle for the expansion of God's kingdom.

It was thus quite a surprise to the man that he was promptly devoured by sharks in the infested area of the ocean where he had piously remained adrift.

Upon entering the afterlife, he quickly made his way to where people in heaven speak directly to God, which we'll assume is some sort of large house with a walkway made of gold bricks leading to its entrance.

He fell to his knees upon being in the presence of the Lord, thanking God for allowing him into His kingdom, but he did have a question about his demise.

"Why, Lord? Why didn't you rescue me? I was faithful! I believed your miracles would save me! Why was my faith unrewarded? Why didn't you come through for me?"

God's booming voice came down on the man, and He said, "Verily, I sent you a boat no less than three times and even sent you a helicopter with a ladder. What part of rescuing you were you not comprehending?"

The point of this modern-day parable may differ based on who is telling it, but for the purposes of this book it is to illustrate how Christians often don't listen to God, because they've decided for Him what He must do and how He must do it. They don't recognize His victory when it appears.

Stop Leaning on Your Own Understanding

God will come through. More specifically, God will come through as I expect Him to.

You've probably heard it a thousand times, perhaps not even fully realizing that's what the person is doing. And you've probably done it yourself.

Very often, when people say, "God will come through," or the old, "God works all things for good," they've actually already decided what options God has to work from.

A relationship you were happy with ends when you find out you were the only one being monogamous? God will come through, and this will lead you to the person you should have always been with.

Lost a job? When one door closes, another will open. God will come through and has your dream job waiting for you behind door number 3.

Failed a major exam and got rejected by your college of choice? Clearly, God wanted you to fail this test so you could go to the college you were always supposed to!

We are constantly committing to God the power over our futures while simultaneously feeling certain we know what that future is. "You drive, but take the next left, though, and slow down."

Throughout my life I have found myself setting the rules for God while believing I have somehow surrendered to Him.

"Lord, I want you to guide me and light my path," I pray. "Please choose option A or option B, for your Glory."

This is dumb.

In actuality, I'm just conforming God to *my* will.

Now, some of the examples I've provided may seem harmless. Many of them may be the types of things you've heard in your life or said to someone else and they actually turned out to be true. That's great!

But that's not contrary information. Just because a few things work out doesn't mean your method is right.

The ultimate lesson, which is relevant as we carry on into discussing Christianity in the Trump era, is that God's plans are not required to conform to what you have decided are your only choices. God's considerations are not held hostage by your desire to solve a problem you have. And importantly, God's goals are always perfect, regardless of any inconvenience it may cause an individual.

All too often, as I've said, we become so devoted to our worldly goals that we find ourselves rationalizing how what we want is also what God wants, regardless of any evidence to the contrary.

We've got goals, and why shouldn't we have them? Doesn't God want us to be happy while we're here? The problem is not that you have goals and that you want God to help you achieve them. That's a fine thing to want, and wanting things is fine. I want ice cream. The problem is also not that you hope and pray for better outcomes related to your goals. Praying is good. There's a whole book about that.

No, the problem comes when you believe that life is about achieving goals. It isn't. I don't mean that "life doesn't include achieving goals," or that "goals are tools of the devil." I said it's wrong to believe that life is *about* achieving goals.

We tend to fall into this trap where we find ourselves believing that life on this mortal coil is something akin to a great television show in which we are the protagonist and God is the wise neighbor whose face is obscured by the fence, and He drops nuggets of wisdom from time to time that help you in your journey toward your goals.

Goals like a big house, a nice job, a fulfilling relationship, and a happy family.

Sure, we always go back to the fence to get those bits of guidance we need from time to time, but for the most part, God simply remains behind that fence. There when we need Him, but not one of the central characters no matter how much we tell others He is.

But we're not the main character. God is. And at the same time, He's the writer. And when we go off script, the whole show falls apart.

Think of the show *Moonlighting* for a moment. The hour-long drama, from the 1980s, followed the exploits of protagonists David Addison (Bruce Willis) and Maddie Hayes (Cybill Shepherd) as private investigators at Blue Moon Detective Agency.

The show had a few years of success and an adoring fan base that tuned in to watch the adventures that unfolded each week, but it was the sexual tension between David and Maddie that kept people on the edge of their seats wondering what would happen next.

Viewers badly wanted David and Maddie to be together. Sure, they often fought and slammed the doors of their offices when they were fed up, but it was all just a denial of their true feelings for each other. Fans were convinced that if the two simply gave in and admitted they wanted to be with each other, they would be so happy and the show would be more fulfilling.

And that's exactly what the writers gave them in season three.

The result? The show was canceled within two seasons.

Debates could be had (and are had in the dark corners of fan message boards), but in general, people agree that the death of the "will they, won't they" chemistry was the death of the show. (Which it obviously was, so just face it and join a new message board.)

When the viewers finally got what they believed they wanted, it left them feeling empty. They had decided that they knew what would make the show better, but they were wrong. It made it worse. In fact, it made it end. Had they not gotten their way, who knows how much longer the show might have stayed on the air.

Expectation, anticipation, cliffhangers, near misses, the pain of unrequited emotion . . . obviously that was what was compelling. This hardly requires explaining.

But we're committed to the *Moonlighting* example now, so let's see it through. What if instead of letting ratings-obsessed executives and goal-oriented viewers dictate the direction of the show, the writers had simply written it how they believed it should unfold? It may have gone on, or even gotten better.

This is the same mistake we make in our lives all the time. We have these goals, and we look at God as little more than an adviser in our pursuit of them, or worse, a crutch or shortcut to achieving them. Eventually, we either succeed in those goals or we endlessly fail to achieve them; in either case, we end up miserable precisely because our life became about those ends. But as I said, life isn't about goals. It's not about our ends.

There is only one goal for a Christian, and that is to surrender the script. To stop trying to write it. To allow God to unfold the story and never assume we know where it's headed. To give ourselves over to Him and let our achievements be a testament to Him so that others might see how surrendering is the only path. God has all the details in His hands, from the job, to the family, to the relationship you are pursuing.

That faithful man floating in the ocean, having already decided how God would perform His work, is all of us at one time or another.

Humans have an incredible ability to rationalize. And when people have decided the script of their lives, it's not hard to see everything as working toward the story they've already written.

So as was mentioned previously, people continue to break their paths down into two options.

Option A: Look past all of the signs that your relationship is toxic and hold out for it to magically change. And then you'll finally be happy.

Option B: Be alone forever.

Option A: Look past all the bad parts of the business plan, mortgage the house, and pile yet more money in until you finally get recognition and wealth for this great idea. And then you'll finally be happy.

Option B: Give up, lose all the hard work you've put in, get a job at a gas station, and be a broken failure forever.

The behaviors that many Christians displayed in their support of Donald J. Trump as a candidate for president and the behaviors they've displayed in his defense since his election fall right into this same dynamic.

Option A: Put all of your effort into supporting a politician who has shown his character to be lacking but who promises to do the things you've been waiting for a president to do.

Option B: Hillary Clinton wins and puts you in a death camp.

In each of these examples, there is a hubris of sorts. A hubris that is harder to detect because, most times, people seem to genuinely believe they are giving their decisions to God.

Any goal on Earth, when compared to the eternity of heaven, is comparatively short.

Granting that, let's accept that there are three categories of goals.

1. Short-term goal (wanting to have sex with someone I just met)
2. Long-term goal (recognizing that it might be best for a potential relationship not to have sex right away)
3. Eternal goal (God has instructed me that my eternal soul is best served by waiting until marriage to have sex at all)

For a Christian, it is important to accept that long-term earthly goals can and do exist, whether or not God is a consideration. The existence of God alters the equation of any goal by adding eternal implications to the process.

Certainly, a short-term goal can be achieved without disrupting the long-term or eternal goals (I'm going to take a break from work because I feel tired), and a long-term goal can be prudent without considering the eternal implications (I'm going to take this job because I'll make more money and live more comfortably), but each does have the ability to disrupt the other. Eternal goals are the only type of the three categories that will never *corrupt* the other two.

With that hierarchy established, let's consider the cost of orienting your life around short-term goals in a purely worldly context.

I have a joke about the nightly struggle of being a single parent making lunches for my kids to take to school:

five p.m.: "The kids should make their lunches for tomorrow."
seven p.m.: "The kids should make their lunches before bed."
nine p.m.: "I should make their lunches now so I'm not rushing in the morning."
six a.m.: "Crap, I've gotta make their lunches."
eight a.m.: "Technically, Lunchables are healthy."

It's funny (and sadly true), but it's a perfect example of how short-term thinking hurts your long-term goals.

At five p.m. I'm usually still working, and the kids are quietly distracting themselves. Getting them to do what they should in that moment would be the responsible move, but it would require effort that would take me away from my work, and I'd really like to be finished so I can relax.

At seven p.m., I'm finally doing that relaxing I mentioned. I really want to enjoy the benefits of it, and we can deal with the lunches, I don't know . . . later.

And so on, and so on. There's always a specific reason that whatever is taking place in the moment is more important than the cost in the future. You may recognize this as basic procrastination.

But by putting off the relatively minor stress of dealing with the lunches when it was most responsible to deal with them (at five p.m.), I've caused myself more problems than would have been had by stopping working early to deal with it.

First of all, I extended my stress since, as is clear from how many times the thought occurred to me, the need to do this was on my mind all night and into the morning, making me anxious. That "relaxation" I sought was completely disrupted by the fact that I chose to wait. I thoroughly undid my entire goal by trying to escape short-term consequences (stopping work early) at the cost of my short-term desires (being able to relax without obsessing over sandwiches) and ultimately hindered my long-term goal (being a good parent).

My choices disrupted several of my parental goals—such as giving my children healthy food for their growing bodies and teaching the responsibility of preparing for the next school day—all in favor of the immediate gratification of feeling less stress about work by accomplishing more of it, and later the simple joy of watching television.

Accomplishing more work can certainly have long-term benefits, and relaxing is absolutely something all people must learn to do if they don't want to die of a heart attack which, I would think, is another worthy goal, but I was under no obligation to sacrifice other long-term goals in order to maintain a healthy work life or the opportunity to relax.

Had I just gotten the children to make their sandwiches at five p.m., I could have returned to work and finished up what I was in the middle of without the mixture of guilt that comes with choosing work over parenting. I could have been better situated to relax later because my work would have felt more fulfilling, and I wouldn't have other responsibilities that I had neglected weighing on me.

Everything—the parental goals, the work goals, the relaxation goals—suffered as a result of my short-term thinking. And in the end, getting Lunchables was the "easy button" solution that I rationalized would make it all okay. "These are healthy enough," I told myself. And by then, having taught myself all the wrong lessons, by using the "easy button" and spending unnecessary money on less healthy solutions to half-ass repair a situation I caused, I managed to make myself feel okay. At the very least okay enough to survive to the next mess I create.

"Screw it," I thought. "It's better than no lunch at all."

Sometimes (and this is, sadly, not an imaginary parable I'm providing) I would have been so frustrated at all the stress I caused myself by putting off such a relatively simple responsibility, I would load up on Lunchables for the week once I got to the store for my emergency solution.

I was literally planning ahead to ensure that I could take the short-term option. I had no confidence I would actually do what I should. Rather than address or embrace the problems associated with doing the right thing every day, I adopted a solution to ensure I could continue to merely survive with subpar outcomes.

To put it more succinctly: the only long-term goal setting I actually managed to pull off in this entire Lunchables saga was protecting my ability to focus on the short term.

That's exactly how modern political life operates, and it is precisely why things are as bad as they are. Short-term thinking is vital to the success of our modern political parties.

Take the Tax Cuts and Jobs Act of 2017, for example. Once the Republican-led legislation was passed and signed into law by President Trump, Americans immediately saw the results in their paychecks. This was a result of the IRS issuing new guidance to employers about how much should be withheld from employee paychecks to stay in compliance with the new tax law.

The Treasury Department said that more than 90 percent of Americans would see more money in their paychecks, which, you could probably guess, people were happy about.[2]

But passing the bill, even with Republicans holding majorities in both the House and the Senate, required compromise, and part of that compromise was that these decreases have an expiration date, phasing out in 2025.

This was not because it's impossible to make tax rates permanent. After all, the corporate tax cuts in the same bill are permanent, so clearly it can be done. So why make them temporary?

There have been scores of policy papers and analyses that have broken down the various reasons and how it would work and why it makes sense, and this book isn't about tax cuts, so let me make this simple.

It is temporary because short-term thinking is how you win in politics. Everything is about the leverage that a party might need later, and in this case, by making them temporary, Republicans ensured that they have a fantastic campaigning tool in the 2024 election.

In other words, the only long-term goal the Republicans went to great lengths to protect was making sure that later on they could once again appeal to the American people's short-term concerns.

Who knows the conditions of our national debt or deficits by 2024? We could be in a budget crisis by then with Democrats calling for taxes to go up to close the gap. There's really no way to know. But what we can know for sure is that the 90 percent of Americans who had spent the previous seven years enjoying more money in their paychecks are likely not going to be too keen on voting for whichever presidential candidate is calling to change that.

The cycle of short-term thinking affects everything. The politicians are thinking only about the short-term goals of acquiring votes and supporters. The voters are thinking only about the short-term goals that are being dangled in front of them. And the long-term health of the country declines because everyone everywhere is just buying Lunchables and telling themselves they'll start making healthier food "tomorrow."

None of this is surprising or shocking information to just about anyone reading this book, but where this plays into American Christianity is of vital importance.

In fact, short-term goals are why Donald Trump is president today.

The Means

The crossroads where evangelicals find themselves is one that many seem unwilling to admit exists.

By employing pseudo-biblical defenses of choosing lesser evils or "accepting our imperfect world" as a way to excuse looking past a powerful figure whose policy agenda you find mostly agreeable, Trump evangelicals are truly making the largest of bets.

They are betting their entire reputation on the possibility that the public, which evangelicals are quick to point out has been distancing itself from the values of Christianity for decades, will somehow see the fruits of their labors and accept that the means that produced them were worth it in service of this greater good they'd ushered in.

It reminds me of Jaime Lannister, eldest son of Tywin Lannister on the immensely popular HBO television series *Game of Thrones*. When Olenna Tyrell, another powerful figure in the medieval world that serves as the show's setting, discusses Jaime's incestuous devotion to his sister, Queen Cersei, she tries to impress upon him that his devotion is misplaced and that victory in the ensuing wars will not earn the reward he hopes, saying "She's a monster. You do know that?"

Jaime responds, "To you, I'm sure. To others as well. But after we've won, and there's no one left to oppose us, when people are living peacefully in the world she built, do you really think they'll wring their hands over the way she built it?" But he looks anxious, as if he doesn't quite believe it himself.

This, above all, seems to be the guiding philosophy that evangelicals have undertaken. It is expressed over and over throughout the quotes I've referenced in these pages.

Belief in "winning at any cost" is a doomed philosophy. One which even a fictional character's facial expressions reveal an unease with saying out loud.

It's something I'd imagine any Christian Trump supporter struggles with at times. Perhaps privately or only among those they trust, but a lie that I can't imagine they are capable of keeping from themselves.

There is no biblical accounting of the means to a good end being a secondary consideration. In fact, in Romans 3:8, when Paul is accused of arguing that we should "do evil that good may come," he calls it slander, saying these accusers' "condemnation is just." Even most secular fiction, from any of history's greatest storytellers, Homer to Shakespeare to Tolkien or Rowling, would not earnestly present the idea that pursuit of a "good end" without consideration for the path that brings you there is a wise course.

I won't attempt to recount all of human history and understanding in one section of one chapter of my book, but all the same, I don't have to. The simplest of wisdoms is that "the ends don't justify the means." I know it to be true, Trump's detractors know it to be true, as do Trump's supporters.

But this self-lie is especially inadvisable for a Christian. It is the hijacking of God's will. It is the belief that God needs *you* to fulfill

His purposes. And it is the antithesis of all biblical teachings I've ever known.

Take the Bill Clinton scandal, for example.

As I laid out in chapter 3, evangelicals were fighting against moral relativism as offered by Clinton's supporters, and the claim those evangelicals made was that there would be a cost to their short-term thinking.

They were right.

Even as recently as two years ago, Democrats continued to rationalize Clinton's behaviors as not reflective of any larger concern than the man's own marriage.

But after the explosion of the #MeToo moment, in which women across the country came forward and exposed the underbelly of a misogynistic culture that had perhaps partly been enabled by situations similar to the dismissals of Clinton's many accusers, some began the difficult process of reflecting on these decisions as a mistake.

Democratic senator Kirsten Gillibrand of New York was quoted in November 2017 as saying that Bill Clinton should have resigned amid the scandal. She clarified when pressed by the *New York Times*,[3] saying, "Things have changed today, and I think under those circumstances there should be a very different reaction," indicating that she might give some grace to decisions made at the time as having been in a different "era."

At Vox.com in late 2017, liberal columnist and Hillary Clinton supporter Matt Yglesias wrote an article titled "Bill Clinton Should Have Resigned."

In it, Yglesias painstakingly makes the case that by excusing or rationalizing President Clinton's behavior, his supporters made light of what was a classic case of a powerful man abusing the vulnerability of a subordinate for sexual gratification and proposed that this was inconsistent with the left's stated values of empowering women and supporting victims.

> *The wrongdoing at issue was never just a private matter for the Clinton family; it was a high-profile exemplar of a widespread social problem: men's abuse of workplace power for sexual gain. It was and is a striking example of a genre of misconduct that society has a strong interest in stamping out. That alone should have been enough to have pressured Clinton out of office.*

While Yglesias's focus is solely on the aspect of how it diminished the moral authority of the left as it relates to this specific issue, the thrust of his point is completely in line with James 1:1–12, expressing the value of high moral expectations for leaders.

At the *Atlantic*, columnist Caitlin Flanagan went further, saying the "Democratic Party needs to make its own reckoning of the way it protected Bill Clinton. The party needs to come to terms with the fact that it was so enraptured by their brilliant, Big Dog president and his stunning string of progressive accomplishments that it abandoned some of its central principles. The party was on the wrong side of history, and there are consequences for that. Yet expedience is not the only reason to make this public accounting. If it is possible for politics and moral behavior to coexist, then this grave wrong needs to be acknowledged."

It took twenty years, and some of the reasoning is certainly different, but the conclusions, while not universal, are starting to reveal a unified interpretation of what would have been "right" in the Clinton era: that there was and perhaps continues to be a cultural reckoning as a result of excusing the actions of Bill Clinton.

Years later the scandal is still referenced often, playing an enormous role in the 2016 presidential election in which Hillary Clinton was the Democratic nominee for president. As his wife, she was portrayed by the opposition as complicit in the attempts to cover up the truth of the Lewinsky affair even as the scandal unfolded.

Once the country had accepted a philanderer in office, reports of Donald Trump's own sexual proclivities, and, at times, open endorsement of adultery, just did not impact a desensitized public in the way they may have prior to the Clinton scandals.

Indeed, if I might speculate, the lowering of expectations that allowed for rationalizations of Bill Clinton's transgressions may very well have played a role in preventing Hillary from seizing the office two decades later.

For many conservatives, the entire Bill Clinton era was a culture war. One that saw Christian values sliding perilously close to the abyss, but which many evangelicals believed could serve as the catalyst for the "spiritual awakening" James Dobson had called for.

As a believer, this is a complicated but vital aspect of trust in God's plans.

Our job is to ensure our devotion to Christ's teachings in the means. Our trust is to believe in God's will as it relates to the ends.

In 2 Chronicles, chapter 16, the king of Judah, Asa, was a follower of God. However, Asa had begun to allow a focus on "good ends," ends he'd determined, to diminish his reliance on God's expectations, leaning instead on shady maneuvering and what he perceived as more expedient methods.

[1] In the thirty-sixth year of Asa's reign Baasha king of Israel went up against Judah and fortified Ramah to prevent anyone from leaving or entering the territory of Asa king of Judah.

[2] Asa then took the silver and gold out of the treasuries of the Lord's temple and of his own palace and sent it to Ben-Hadad king of Aram, who was ruling in Damascus. [3] "Let there be a

treaty between me and you," he said, "as there was between my father and your father. See, I am sending you silver and gold. Now break your treaty with Baasha king of Israel so he will withdraw from me."

4   Ben-Hadad agreed with King Asa and sent the commanders of his forces against the towns of Israel. They conquered Ijon, Dan, Abel Maim and all the store cities of Naphtali. 5 When Baasha heard this, he stopped building Ramah and abandoned his work. 6 Then King Asa brought all the men of Judah, and they carried away from Ramah the stones and timber Baasha had been using. With them he built up Geba and Mizpah.

In the context of the era, this was brilliant and pragmatic maneuvering on Asa's part. His enemy defeated, Asa actually used the resources of that defeated foe to build two new cities. There's little doubt that his subjects saw this as a welcome outcome to Asa's efforts. It could even be objectively stated that this was a "good" outcome.

And yet, God was displeased with Asa.

7   At that time Hanani the seer came to Asa king of Judah and said to him: "Because you relied on the king of Aram and not on the Lord your God, the army of the king of Aram has escaped from your hand. 8 Were not the Cushites and Libyans a mighty army with great numbers of chariots and horsemen? Yet when you relied on the Lord, he delivered them into your hand. 9 For the eyes of the Lord range throughout the earth to strengthen those whose hearts are fully committed to him. You have done a foolish thing, and from now on you will be at war."

Essentially, God was angry with Asa for appealing to the king of Syria's sinful greed and exploiting his susceptibility to that temptation. No matter the outcome, God judged Asa's actions immoral, and, as the seer pointed out, this was after God had already demonstrated why Asa should trust Him, when despite the overwhelming odds Asa had faced against the Cushites and the Libyans, God provided the means to victory.

It must have been quite easy for Asa to look at the opportunity in front of him and the ease with which he could manipulate his enemy to become his ally. He may have known on some level perhaps that this "easier" route would not be something he could imagine God asking him to do. And yet it all seemed so obvious to him.

This concept, which certainly has biblical basis, is commonly accepted outside of the confines of religion. Comedian Louis C.K. perfectly describes the paradox one confronts when faced with the "easy way" not being the right way and how to do the right thing, one must simply prioritize a higher standard than the preference of their self-interest.

> My nine-year-old, she's just figuring out about lying and that's a tough thing. It's hard to roll that one back, because lying is pretty amazingly useful in life. It's like, how do you tell a kid not to use a thing that just solves every possible problem, like magic?
>
> But to a little kid, trouble is like this horrible . . . Did you take the chocolate? And she did and she doesn't know how to handle it. Did you— Did you take it? "No." Well, all right, then, have a nice day. How do you then tell her, yeah, don't ever apply that perfect solution again, to terrifying things?[4]

Somewhat poetically, the lesson he is very adeptly teaching is one he had to publicly reckon with after several women accused him of deplorable sexual misconduct,[5] causing him to cancel all public appearances and go into self-imposed exile from the spotlight following an apology for his actions.

In the apology, some of his comments highlight exactly why we teach our kids not to lie. "At the time, I said to myself that what I did was okay," Louis C.K. said, indicating that on some level, he was aware that using his influential position to achieve immediate sexual gratification was wrong, but that his desire for this short-term benefit superseded his moral compass, and he rationalized to himself that it was okay because the women gave consent despite it being under what could be described as reputational duress.

C.K. added, "I have been remorseful of my actions. And I've tried to learn from them. And run from them."

While we may not all be guilty of the same types of transgressions as C.K. or Bill Clinton, in all likelihood we have found ourselves in a position to rationalize a bad course because we see a desirable end and a path to it.

In these situations, the short-term benefit should always take a back seat to a higher standard. Because, as the Bible and much of our collective life experiences have taught us, there is and always will be a cost. And often that cost dismantles the very good end you thought you'd achieved.

In the case of Louis C.K., it cost him his career.

Furthermore, let's accept for a moment that this "new world" Trump's chorus of apologists continue to claim to be working toward will ultimately be realized.

What exactly would Christians' "ends justify the means" strategy teach the nonbeliever? What would they learn from it about how the world operates according to God's design?

Would they have learned that there is reward for remaining principled? No, of course not. Quite the opposite. They would have learned that principles are flexible so long as the end goal is clear and "good."

Would they have learned that there is a consequence for lack of repentance? Again, clearly not. Trump, even among his most ardent

Christian supporters, has never been credibly described as penitent. On its own, this could be considered a private struggle between Trump and God, at least were it not for the minor point that his supporters have taken it upon themselves to defend and support this impenitence, calling it virtuous. And they do this in full view of a world that already harbors doubts as to the sincerity of evangelical devotion to biblical principles. To make matters worse, as I've demonstrated, in the years since his election many of these same supporters, as well as many of his most influential allies, have adopted this same reckless impenitence for themselves.

Would they have learned the importance of humility? This would be hard to accomplish when Trump—who even his fans admit serves as the poster child for narcissism—is applauded at every turn by his supporters, including evangelicals, who are quick to congratulate him on his forthrightness and lament that other leaders aren't as brash. His supporters have even declared war on Republicans who won't employ the same unapologetic narcissism, threatening to primary senators like Ben Sasse of Nebraska for not joining in on the celebration of arrogance. Again, Trump being who he is is not nearly as damaging as the rationalizations his supporters use to excuse or celebrate what should be considered negative or worthy of criticism.

Would they have seen the compassion of Christ as expressed through His followers? Not likely. "Compassion" has mostly taken a back seat in favor of a dispassionate form of "pragmatism" that, from most accounts I've seen, puts the ownership of America's possessions and autonomy at a much higher value than the well-being of those who may suffer elsewhere.

So what would those outside of the evangelical movement have seen then in this hypothetical Trump-created conservative utopia? What would they have learned of God's purposes for humanity?

They would see a man celebrated not for his accomplishments as a president but primarily for characteristics that historically defined the antithesis of the Christian ideal. He is a man driven openly by desires for wealth, power, lust, and stature. Upon being lifted to the pinnacle of Western civilization, he showed little or no regard for the values that Christ told us to look for in leaders, and he managed to achieve this position thanks to his adoring supporters, who reveled in the most troublesome aspects of his character.

They would learn a very simple lesson. That America doesn't need God. And they certainly don't need His witnesses either. The Bible is a lie.

But, hey, at least we can say "Merry Christmas" again without being ashamed.

AFTERWORD:
LOVE YOUR ENEMIES

Even after all that you've read, you may still be wondering what precisely it is that I intended in naming this book *The Immoral Majority*.

The quick and easy answer is that the title is a protest of sorts. It is intended to offer a contrast to the preeminent evangelical group of the last fifty years, the Moral Majority, an organization whose name, to many, implied a sense of exclusion. As if they were saying "we" are moral and "they" are not. Since I have been making the case that the approach of that self-same group is rooted in misplaced devotion—idolatry, even—then *The Immoral Majority* felt like the obvious title.

But I'd argue that my title does not suffer from the same exclusionary connotation as the late Jerry Falwell's organization. Far from being a judgment targeted specifically at evangelicals—despite the fact that in these pages I have singled that group out for criticism—the Immoral Majority is a group in which we can all claim membership. Every flawed human born in all of history. We all fall short, we all fail, and we are all guilty.

Of course, this raises a different question, one that sounds pedantic but, in reality, is of serious import to the underlying thesis of this book.

"If the Immoral Majority includes every human ever born, then shouldn't it be called the Immoral Totality?"

The Bible gives us the answer.

1 Peter 2:21-24:

> For to this you were called, because Christ also suffered for us, leaving us an example, that you should follow His steps: "Who

committed no sin, nor was deceit found in His mouth"; who, when He was reviled, did not revile in return; when He suffered, He did not threaten, but committed Himself to Him who judges righteously; who Himself bore our sins in His own body on the tree, that we, having died to sins, might live for righteousness— by whose stripes you were healed.

Jesus lived the perfect life and, as God made flesh, was the perfect man. But make no mistake, He was a man. Like us, but apart from us. Not with us, but for us. Yet a man all the same. He occupies the singular minority position in all of history as the only *inherently* moral human.

In other words, there *is* no moral majority. It was always a misnomer. Apart from Jesus, there is us. The rest of us. The immoral majority.

And in case I'm not being clear enough, let me specify. It's not just Christian conservatives. It is also those who rejected them in the first place. We all have uncomfortable questions to ask ourselves in the wake of this immoral Trumpian movement.

That's not a pat answer. In a very real political sense, the modern conservative Trumpist right was the direct electoral result of the previous decades of politics. The polarized Obama years, the excesses of political correctness culture, the hyperbolic liberal claim that most things are motivated by racial hate. We have examined at length here the many morally bankrupt ways to respond to those things, but that does not absolve those things of responsibility.

It takes two to Trumpo.

The mass media and popular culture vilify things that do not deserve vilification. Hollywood does not broadly mock religion but specifically mocks Christians on the very premises of Christian faith itself. From the teachings of Jesus to the Virgin Birth to the very creation of the universe, popular secular culture is a condescending and bile-soaked assault on the values of everyday Americans, and has been for decades now.

Likewise, the economic suffering in small-town America is real and hard and exacts a heavy price. The opioid crisis, the war on drugs, the stock market failures, the crashes, the bubbles, the tech industry overall . . . all of this cannot be free of blame just because a segment of the voting population has failed to respond to those challenges gracefully.

A person may act a coward in the face of disaster, but that doesn't mean they caused the disaster. Or even deserved it. It just doesn't. Morality doesn't work that way. Morality is not situational. The same standard applies to Christians and conservatives as to liberals as to secularists.

In *Mere Christianity*, C. S. Lewis talks at length about what he calls the Law of Human Nature.

> *These, then, are the two points I wanted to make. First, that human beings, all over the earth, have this curious idea that they ought to behave in a certain way, and cannot really get rid of it. Secondly, that they do not in fact behave in that way. They know the Law of Nature; they break it. These two facts are the foundation of all clear thinking about ourselves and the universe we live in.*[1]

What Lewis is saying is that people, both theist and anti-theist, tend to know there is a standard for what is right and what is wrong. That that standard is apart from themselves (not inherent or unique to them) and that, in general, people tend either to fall short of upholding it, or simply to reject it altogether.

They believe that something can be ultimately right or wrong, they believe they know which things are which, and then they do the wrong things anyway.

A Christian would call it our fallen nature. A secularist might call it the human condition. But the idea is the same. Aside from the decision to be wrong, the more interesting aspect is the failure to be right. It is in those who know and believe in the concept and strive to be good and better but consistently fail to do so. Mankind broadly attempts to improve, to be just and moral and righteous, and yet, with utter consistency, falters, falls short. Man tries. Man fails.

Be wary of any Christians who claim they command an inherent grasp of the moral authority that might guide someone away from those failures. They have surely failed to understand some core principles that the Bible teaches, since, as I've said, Christians are not tasked with looking to ourselves for answers. As G. K. Chesterton wrote in *Orthodoxy*, "A man was meant to be doubtful about himself, but undoubting about the truth." Christians are tasked with sharing the authority of God. All of our judgments of right and wrong are supposed to be derivative of that understanding.

In Christianity, it is not simply that His word is the law. He *is* the law. These things that God says to look to Him to understand are precisely because His authority is not a result of what He "does." They are the manifestation of His perfect and unchangeable nature.

Therefore, we are all flawed. Does that mean, given our own imperfection, that we can make no valid moral critiques of others? In no readings of the Bible have I ever come to the conclusion that God wishes us to abdicate moral discernment.

Quite the opposite, really.

The verse most often used to give the false impression that God commands us to "live and let live" is also one of the most misinterpreted verses of the entire Bible.

MATTHEW 7:1–3

1 "Do not judge, or you too will be judged. 2 For in the same way
   you judge others, you will be judged, and with the measure you
   use, it will be measured to you.
3 "Why do you look at the speck of sawdust in your brother's eye
   and pay no attention to the plank in your own eye?"

A valued dictum of the theological left, and more recently the
conservative right, this verse is often used to excuse someone from
consequence or responsibility for their actions. It's the "no labels"
of religious quotes. This is a fundamental misreading of the scrip-
ture.

Other passages that support the actual intent of this verse are
not difficult to find. God clearly lays out a moral imperative to
make judgments about others' character.

LEVITICUS 19:15–18

15 Do not pervert justice; do not show partiality to the poor or
    favoritism to the great, but judge your neighbor fairly.
16 Do not go about spreading slander among your people.
    Do not do anything that endangers your neighbor's life. I am the
    Lord.
17 Do not hate a fellow Israelite in your heart. Rebuke your
    neighbor frankly so you will not share in their guilt.
18 Do not seek revenge or bear a grudge against anyone among your
    people, but love your neighbor as yourself. I am the Lord.

Taking these verses together, the truth is much more nuanced
than "don't judge." We are actually *called upon to* judge others, especially those around us who also call themselves Christians.
However, we are not to judge that over which God alone has

dominion, such as a person's soul, but to judge and operate in the context of our sphere of discernment: recognizing right and wrong and speaking loudly when confronted with either.

The key, as is clearly laid out in these passages, is that God's authority is the source of the judgment ("I am the Lord"), and that those leveling the charge must themselves be in a position to discern without the taint of their hypocrisy diminishing God's authority.

So does this mean that only perfect people are allowed to offer judgment or call something immoral that is, in fact, immoral?

Certainly not. If this was the case, only Jesus himself could follow His own teachings. With the exception of Jesus, all persons in the history of humanity who have brought God's morality to another have themselves been an immoral vessel of that truth.

In all things we must be consistent.

If I were to tell someone that stealing is wrong as I am stealing, I am not showing consistency. I'm not showing an awareness of the log in my eye. Even if I know in my mind that stealing is wrong and I'm a hypocrite for admonishing someone else while I engage in the same activity, I have sullied the truth of God's morality by allowing my sin to interfere with the enlightenment of the person I'm condemning. In this example, the fact that I have not offered any intent to hold myself to the standard to which I am holding them will undoubtedly interfere with their enlightenment. Actions speak louder than words.

If I were stealing while telling someone else stealing is wrong, acknowledging my hypocrisy even as I commit it, I've still not separated my sin from bringing God's truth to the matter.

However, none of this necessarily prevents a thief from telling someone stealing is wrong.

The question for that thief to ensure the consistency of his position is simply, "Why then have you stolen?" And the answer is, "Because I, like you, am a sinner. And I endeavor to sin no more."

With genuine repentance of your wrongs, you can offer God's truth, regardless of your previous sins. In fact, oftentimes having an awareness of your sins while showing a commitment to resist them can be a powerful message to someone who struggles with similar temptations.

Making it personal for a moment, my Twitter feed is again a stellar example of this. As I've stated, it's often littered with arguments and insults, curse words and sarcasm that is not becoming of a man seeking God's command that we use caution and restraint in our words.

It may seem like a petty point to some, but the fact is, I recognize this log in my eye. And while I know that I must work to be better and that I must find a way to remove that log, I try not to bring judgment to others who commit the sins I commit in this regard because to do so would sully the truth of God's moral perfection. I would be bringing my sinful nature to His judgment, and I would potentially hinder the person I'm attempting to help.

Does my blatant sinning in some areas of my life prevent me from bringing God's moral clarity to another situation? No. An adulterer is completely capable of telling someone else murder is wrong without at all impugning the righteousness of that judgment.

The nuance of this begins to develop as a result of an overall picture one gives of oneself. If someone continues to sin unrepentantly while claiming to be a Christian, it is fair for another Christian who is positioned to admonish them, to offer judgment on those behaviors. But whoever it is that offers this admonishment needs to be sure they've given themselves a thorough look in the mirror and addressed their potential hypocrisies.

As for me, I have made it a personal conviction never to question the sincerity of another person's faith. Many Christians have openly mocked the idea that President Barack Obama, and later, President Donald Trump, are Christians, both of whom claim they are. Perhaps someone is in a position to question these things, but it is certainly not me.

I have serious political and philosophical differences with both men, but I've never felt that I am in the position to question their relationship with God. I have admonished several things that each has individually done that I find to be not in keeping with God's wisdom, but to bring the entirety of that relationship with God into question is simply a bridge too far for me, given that my own public discourse at times leaves much to be desired in terms of projecting the Christianity I should project. To put it simply: the log is too big for me to offer judgment of others.

That brings me back to the evangelical movement at large, which has a rather massive log in its eye at the moment. For the possibility of a bit of worldly influence, they surrendered their moral voice in the public sphere.

We discussed in the previous chapter that there is a hierarchy to choosing which moral laws are of greater priority than others. It is in this that I can most easily summarize how the evangelical movement that has embraced Trumpism has failed. There is no question as to what the highest moral laws are since Jesus was specifically asked in Mark, Chapter 12.

²⁸ One of the teachers of the law came and heard them debating. Noticing that Jesus had given them a good answer, he asked him, "Of all the commandments, which is the most important?"

²⁹ "The most important one," answered Jesus, "is this: 'Hear, O Israel: The Lord our God, the Lord is one. ³⁰ Love the Lord your

God with all your heart and with all your soul and with all your mind and with all your strength.' [31] The second is this: 'Love your neighbor as yourself.' There is no commandment greater than these."

It is by comparing evangelicals' track record with the standard of the highest law that their failure becomes clear. Because it really doesn't matter who you voted for. The problem with the "lesser evils" principle isn't that we can never choose between bad options—it's that in that dilemma the *motivation* for choice is framed as mere practicality, not prayerful obedience. Did your conscience tell you to vote for Donald Trump? Or for Hillary Clinton? Or not to vote at all? If you follow your conscience, you have made a righteous choice.

The important choice was never between Donald Trump and Hillary Clinton—the important choice was between self-interest and the idolization of "winning" versus loving God and one another. And as I've demonstrated throughout these pages, far too many evangelicals have chosen the former over the latter.

Bitterness over faith. Vengeance over justice. The world over the soul.

This has been the story of the evangelical movement embracing someone who was not in it or of it, who was not like it and did not like it, and who represented culturally and morally all that it opposed. This seemed unlikely, until it became evident how many things the evangelical movement in America already held as values that were not particularly biblically based.

The evangelical Trump voters set aside their religiosity and their moral high ground in favor of winning elections, exacting vengeance, proving a point, defying a stereotype, and protecting the culture as they saw it.

This is the story of evolution, ironically. The movement evolved, or its adherents did. It is an understandable, if not always justifiable, journey. It is a path we can retrace. Steps we can go back over. There were leaders and moments of clarity.

What we know now is that the American Christian conservative is no longer defined the way it once was. We find now a Christian movement that is utterly unfazed by its political savior committing adultery, paying for his mistress's silence through questionable means, and lying about it to everyone, including those selfsame Christian voters.

It has become a movement that believes the role model for their own lives and the upbringing of their children should be defined by their willingness to embrace brutality over compassion, self-interest over persuasion, and deception over repentance.

This is the new evangelical movement in America. In a very real sense, they are evangelizing a rejection of their own purported principles, and yet are unwilling *and* unable to see it. They see no contradiction. The cognitive dissonance is as utter as it is widespread. A unified and mass self-deception.

That is the story you have heard. It is a sad one. It is the story of worldly gain at the cost of the soul, and it is ongoing.

Evangelical Trump voters will despise this book. They will call it a betrayal, a selling out to the left for "pats on the head" or invitations to parties, but that's not really why they'll hate it. They will hate it because, for many, it exposes them.

Not to you, you understand, or even to me, or even, tellingly, to God. No, because it exposes them to themselves. This is a movement that despises little else more than a mirror.

What you can take away, what you can learn is not to hate Christians or religion or faith or the faithful. Not even to hate evangelicals or even Trump evangelicals.

You may choose to pity them, but better to mourn them, as I do, for what was lost was something that may have been of value and worthy of admiration. But even that is merely something you can do or feel, it's not what to take away. It's not the thing to learn.

The thing you can learn, if you are an outsider reading this, is about yourself. Because after all this recrimination and blame and harsh observation, you may have to face the reality of where you stood in every stage of this evolution, and I hope you will.

I have.

Perhaps it was you they faced off against. Perhaps it was you they were angered by and resentful of, you who left them behind and left their small towns in ruin and failed to reach out to them, you who spent so long attacking people of simple values and simple faith as people to be despised only because you found their simplistic views so mundane.

And now, after having been treated despicably, they have embraced the designation "deplorable" and proudly profess their despicable actions while worshipping at the altar of winning. They have become their full deplorable selves.

Was that you who treated them this way? And in this case, where the failure of their own character is on them, your failure was the indifference that caused you not to help someone else as they fell.

If you're honest with yourself, I think you know that part of the "solution" among the Trump evangelical's panel of critics is that we are simply waiting for them to die. Waiting for their ideology and their character to be buried with them.

But that will not be the story. And in all human history it never has been.

Division is a path that we as a nation have undertaken. Paradoxically, division requires unity. The worst kind of unity. And we have all participated. It is only together that that might change.

The solutions cannot and will not be found to the exclusion of those we see as our enemy.

If you wish to be all that Donald Trump and his ilk are not, then the greatest service you could do for the world is to love them despite themselves. Love doesn't require agreement. It doesn't require compromise. It doesn't require surrender or the shedding of values.

It only and ever required the simple truth that we are stuck together. And if things are going to get better, you cannot wait for others to do it first.

Mankind must improve. It must achieve. It must be good, and to do so it must become better. It must accept that it can fail. Can fall.

This is how we will turn the tide. By recognizing that while we can win, we can lose all the same.

Winning is fleeting. Victory is eternal.

ACKNOWLEDGMENTS

To my brother, Caleb. There's simply no way this book could have been written without you. A far better writer, a far more intelligent analyst, and, unfortunately for you, an excellent editor. Thank you for going through the thousands of words I sent you and offering thoughts and feedback and suggestions and ideas. There are many people to thank, but above all, your presence is felt in this book. I doubt I can repay that assistance with comparable assistance because, unfortunately for you, you're simply much better at this than I am.

To Mom and Dad: I know I already thanked you in the dedication but it's hard to overstate how this book would not have been possible without your support or the foundational beliefs you instilled in me as I grew up. Together, the two of you taught me empathy and reason and, more important, that those two concepts can coexist, which is the basis of this entire book. Thank you.

To my ex-wife, Breeanne: Thank you for all the times you helped lift my burdens with the kids when you knew I needed the extra time. Thank you for letting me walk over to your house in the middle of your workday and read to you to get your thoughts. Thank you for being my target audience and brainstorming thoughts with me when I asked. And thank you for bringing me coffee that day of the deadline. Most important, thank you for the gift of our children and for being my partner in raising them. We've come a long way.

To my daughter, Mia: Thank you for being the inspirational writer you are. Your future is so bright, and I am so excited to have discovered you're much better at this than I'll ever be. Your brilliance and talent cannot be contained, and I cherish every word you write.

To my daughter, Abby: Thank you for your amazing creative mind, your unwavering empathetic spirit, and your uncontainable desire for justice. You are wonderfully complex and, like your siblings, bursting with talent. I can't wait to see what the world has in store for you.

To my son, Colin: Thank you for your inquisitive mind and your patient, loving spirit. You are a born entertainer and analyst, and on top of that you're funny, kind, and eager to make others happy. It's amazing to watch you interact with your friends and see how excited they are to be around you. When you and I watch YouTube theory videos together about the latest film, I constantly marvel at the subtle clues and symbolisms that you pick up and I miss. You, like your sisters, are a born storyteller.

To my daughter, Chloe: Thank you for being the brightest star I've seen. You ooze love and kindness out of every pore of your body. As I wrote this book, I was blessed over and over with your kind notes and your thoughtful gifts. You're hilarious and giving, and want nothing more than to love and be loved. When we read stories together, I'm amazed at what you are able to understand and predict. I know that you, too, have those talents of your siblings.

To Bridget: I am so grateful you were with me toward the end of writing this, as your brain and your perspective were invaluable. You helped me rearrange my thoughts and focus my energy. You did a faux interview with me just so that I could put my thoughts together. You were then kind enough to bring me on your popular podcast, tweet about my book, and overall be a fan of my work, which I'm forever grateful for. You were the first one to get an advance copy of my book and the first one to read it. I could not have finished it without you. Thank you.

To Evan: I've known you since I first started working with you in the church youth group, when you were a rising sixth-grader. Even back then you were already showing a depth and maturity beyond your years. I spent so much time working with you and talking to you and I always wanted to be able to give you pearls of wisdom where I could. Since you've become an adult and we've become better friends, I can't tell you how much I've learned from you. Your insights and our late-night discussions helped me so much in putting together my thoughts for this book. Thank you.

To Fiona: Thank you for being there the day I found out I would be writing this book. And thank you for letting me read to you, and for listening to my original introduction and having the guts to tell me you hated it. Getting genuine feedback from people who care about you isn't always easy, but I knew I could trust you. Thank you for all the times you lifted my burdens with the children you grew to love, and for giving me the space I needed to work when I needed it, even if it meant sacrificing what you wanted. I couldn't have written this without you.

To Jill Jackson: Thank you for more than twenty-five years of friendship. Thank you for being the one person who knew me well enough to know where to push back, when to tell me it was boring, when to smack me around, and when to force a confession of procrastination out of me. Thank you for being my best friend.

To Matt Barcalow: Thank you for the many years as my mentor when we worked with our guys in the youth group. Your honesty, advice, and devotion to seeking Christ have had an enormous influence on my life and this book, and I'll always be grateful that it was you who baptized me all those years ago. Thank you for your friendship and your counsel.

To Lisa: Thank you for the endless availability of discussion and opinion, compromise and disagreement, and bringing me perspectives that I simply couldn't have found elsewhere. You and I come from such different points of view on so many things, yet I never felt that at our core we doubted each other. You brought me invaluable insight and were so kindly available to listen to me ramble on endlessly yet always seemed able to subvert my expectations in how you responded, which was almost always an education. Thank you so much for your help.

To Nicole: Thank you for all the brainstorming sessions, for all the advice, for all the times you brought me stories and ideas, and for being an objective and thoughtful sounding board despite how much I know you struggled to agree with my conclusions.

To my agent, Keith, and my editors, Eric and Hannah: Thank you for your guidance and patience as you walked me through all this. The belief you had in me felt unearned but I'm grateful you saw what was possible.

Also, thank you to Emily Wagner, Chris Loesch, Sarah Smith, Jill Bates, Matt Hilsmier, Dana Loesch, Jon Henke, Lachlan Markay, Casey Knight, Michael Deppisch, Richard Howe, Jay Caruso, Andrea Ruth, David French, Jonah Goldberg, Yoni Appelbaum, Erick Erickson, and Rick Wilson.

And finally, thank you to God. I hope I've made You proud. I hope my ego stayed out of what I wrote and that it is exactly what You'd wish for me to say. Thank You for your gifts, for my children, for my life, and for my soul.

NOTES

INTRODUCTION

1. "Chick-fil-A Donated Nearly $2 Million to Anti-Gay Groups in 2009," *Huffington Post*, December 7, 2017, https://www.huffingtonpost.com /2011/11/01/chick-fil-a-donated-anti-gay-groups-2009_n_1069429 .html.

2. Ben Howe, "Not Without My Chicken," YouTube, August 2, 2012, https:// www.youtube.com/watch?v=jku_4IVJ5ik&app=desktop.

3. Ben Howe (@BenHowe). 2012. "Douchebaggery, thy name is this guy." Twitter, August 2, 2012, 7:12 p.m. https://twitter.com/benhowe/status /230848620075810816?s=21.

4. Ben Howe (@BenHowe). 2012. "We should find out where the guy in this video works." Twitter, August 1, 2012, 7:41 p.m. https://twitter.com/ben howe/status/230855692901302272?s=21.

5. Ben Howe (@BenHowe). 2012. "We didn't get this guy fired." Twitter, August 2, 2012, 3:20 p.m. https://twitter.com/benhowe/status/2311 52417369096194?s=21.

CHAPTER 1: THE SHIFT

1. Jerry Falwell (@JerryFalwellJr). 2016. "Honored to introduce @real DonaldTrump." Twitter, June 21, 2016, 11:39 a.m. https://twitter.com /JerryFalwellJr/status/745325187776811008.

2. "5,000 Attend Anti-Pornography Rally," *San Bernardino Sun*, September 3, 1985, sec. A.

3. Anita Gates, "Tammy Faye Bakker, 65, Emotive Evangelist, Dies," *New York Times*, July 22, 2007, https://www.nytimes.com/2007/07/22/us /22bakker.html?_r=1&.

4. Wayne King, "Swaggart Says He Has Sinned; Will Step Down." *New York Times*, February 22, 1988, https://www.nytimes.com/1988/02/22/us /swaggart-says-he-has-sinned-will-step-down.html.

5. Bruce Buursma, "Swaggart Confesses, Leaves Pulpit—For Now." *Chicago Tribune*, February 22, 1988, https://www.chicagotribune.com/news/ct -xpm-1988-02-22-8804010367-story.html.

6. Frances Frank Marcus, "Swaggart Found Liable for Defaming Minister," *New York Times*, September 13, 1991, https://www.nytimes.com /1991/09/13/us/swaggart-found-liable-for-defaming-minister.html.

7. Rian Dundon, "This Televangelist Cried in Front of 8,000 People After Being Caught with Prostitutes," *Timeline*, May 2, 2018, https://timeline

.com/this-televangelist-cried-in-front-of-8-000-people-after-being-caught
-with-hookers-5fcdde5c6908.

8. *Primetime Live*, "The Apple of God's Eye," produced by Robbie Gordon, aired November 21, 1991, on ABC.

9. Anson D. Shupe, ed., *Wolves Within the Fold: Religious Leadership and Abuses of Power* (New Brunswick, NJ: Rutgers University Press, 1998).

10. Gchinnici1974, "Peter Popoff Exposed—Part 1," YouTube, February 14, 2008, https://www.youtube.com/watch?v=SNl52deOZro.

11. Sarah Pulliam Bailey, "In an Age of Trump and Stormy Daniels, Evangelicals Face Sex Scandals of Their Own," *Washington Post*, March 30, 2018, https://www.washingtonpost.com/news/acts-of-faith/wp/2018/03/30/in-an-age-of-trump-and-stormy-daniels-evangelical-leaders-face-sex-scandals-of-their-own/?utm_term=.612f36eedc21.

12. Art Toalston, "Frank Page Resigns over 'Morally Inappropriate Relationship,'" *Christianity Today*, March 28, 2018, https://www.christianitytoday.com/news/2018/march/frank-page-resigns-southern-baptist-executive-committee-sbc.html.

13. Ed Stetzer, "Andy Savage's Standing Ovation Was Heard Round the World. Because It Was Wrong," *Christianity Today*, January 11, 2018, https://www.christianitytoday.com/edstetzer/2018/january/andy-savages-standing-ovation-was-heard-round-world-because.html.

14. Robert Downen, Lise Olsen, and John Tedesco, "Abuse of Faith: Houston Chronicle, San Antonio Express-News Investigation Reveals Decades of Sexual Abuse from Southern Baptist Church Leaders, Volunteers," *Houston Chronicle*, https://www.houstonchronicle.com/local/investigations/abuse-of-faith/.

15. Ibid.

16. John Tedesco, Lise Olsen, and Robert Downen, "Southern Baptist Leaders Quickly Clear 7 Churches, Sparking Outrage Among Victims," *Houston Chronicle*, February 25, 2019, https://www.houstonchronicle.com/news/houston-texas/houston/article/Southern-Baptist-leaders-quickly-clear-7-13643282.php.

17. Darren Patrick Guerra, "Actually, Most Evangelicals Don't Vote Trump," *Christianity Today*, July 26, 2016, https://www.christianitytoday.com/ct/2016/march-web-only/actually-most-evangelicals-dont-vote-trump.html.

18. Max Lucado, "Max Lucado: Trump Doesn't Pass the Decency Test," *Washington Post*, February 26, 2016, https://www.washingtonpost.com/posteverything/wp/2016/02/26/max-lucado-trump-doesnt-pass-the-decency-test/.

19. Sarah Posner, "Meet the Evangelicals Who Hate Donald Trump," *Rolling Stone*, June 25, 2018, https://www.rollingstone.com/politics/politics-news/meet-the-evangelicals-who-hate-donald-trump-225654/.

20. John Stemberger, "3 Questions Evangelicals Should Ask About Donald Trump—CNNPolitics." CNN, January 6, 2016, https://www.cnn.com/2016/01/05/politics/evangelicals-donald-trump-questions/index.html.

21. Katie Zezima, "'There's Nobody Left': Evangelicals Feel Abandoned by GOP After Trump's Ascent," *Washington Post,* May 8, 2016, https://www.washingtonpost.com/politics/theres-nobody-left-evangelicals-feel-aban doned-by-gop-after-trumps-ascent/2016/05/08/a133991e-130f-11e6 -8967-7ac733c56f12_story.html?utm_term=.a077024870f4.
22. Russell Moore, "Russell Moore: Why This Election Makes Me Hate the Word 'Evangelical.'" *Washington Post.* February 29, 2016, https://www.washingtonpost.com/news/acts-of-faith/wp/2016/02/29/russell-moore -why-this-election-makes-me-hate-the-word-evangelical/?utm_term =.f5116c7913c0.
23. Jessica Taylor, "True Believer? Why Donald Trump Is the Choice of the Religious Right," NPR, September 13, 2015, https://www.npr.org /sections/itsallpolitics/2015/09/13/439833719/true-believer-why-donald -trump-is-the-choice-of-the-religious-right.
24. Donald Trump (@realDonaldTrump). 2016. "Russell Moore is truly a terrible representative." Twitter, May 9, 2016, 3:05 a.m. https://twitter.com /realdonaldtrump/status/729613336191586304?lang=en.
25. Ed Stetzer, "Lord, I Thank Thee That I Am Not Like Those Evangelical Trump Supporters," *Christianity Today,* June 3, 2016, https://www.christianitytoday.com/edstetzer/2016/june/trump-supporters.html.
26. Emma Green, "Trump: 'Your Leaders Are Selling Christianity Down the Tubes,'" *Atlantic,* June 21, 2016, https://www.theatlantic.com/politics /archive/2016/06/trumps-play-to-win-evangelical-voters/488075/.
27. Laurie Goodstein, "Falwell: Blame Abortionists, Feminists and Gays." *The Guardian,* September 19, 2001, https://www.theguardian.com/world /2001/sep/19/september11.usa9.

CHAPTER 2: THE NEW GOOD NEWS

1. Gregory A. Smith and Jessica Martínez, "How the Faithful Voted: A Preliminary 2016 Analysis," Pew Research Center, November 9, 2016, http:// www.pewresearch.org/fact-tank/2016/11/09/how-the-faithful-voted-a -preliminary-2016-analysis.
2. Elizabeth Podrebarac Sciupac and Gregory A. Smith. "How Religious Groups Voted in the Midterm Elections," Pew Research Center, November 7, 2018, http://www.pewresearch.org/fact-tank/2018/11/07/how -religious-groups-voted-in-the-midterm-elections/.
3. Kyle Mantyla, "Jim Bakker: Trump's Election Is 'The Greatest Miracle I Have Ever Seen.'" Right Wing Watch, November 10, 2016, http://www.rightwingwatch.org/post/jim-bakker-trumps-election-is-the-greatest -miracle-i-have-ever-seen/.
4. Lindsay Bever, "Franklin Graham: The Media Didn't Understand the 'God-Factor' in Trump's Win," *Washington Post,* November 10, 2016, https://www.washingtonpost.com/news/acts-of-faith/wp/2016/11/10 /franklin-graham-the-media-didnt-understand-the-god-factor/?utm _term=.ca5b8a49335b.

5. Josiah Ryan, "Trump Fan Compares Trump to Harlots in Bible," CNN, October 10, 2016, https://www.cnn.com/2016/10/10/politics/trump -supporter-god-harlots-bible/index.html.

6. Greg Garrison, "Alabama Pastor Asks Church to Pray for Trump, Against Witchcraft Attacking Him," Al.com, August 22, 2018, https://www.al.com /living/index.ssf/2018/08/alabama_pastor_asks_church_to.html.

7. Tim Hains, "Trump Voter: If Jesus Christ Came Down from the Cross and Told Me Trump Was with Russia, I Wouldn't Believe Him," Real- ClearPolitics, November 21, 2017, https://www.realclearpolitics.com /video/2017/11/21/trump_voter_if_jesus_christ_came_down_from_the _cross_and_told_me_trump_was_with_russia_i_wouldnt_believe _him.html.

8. "About Dr. Jeffress," Pathway to Victory, https://ptv.org/who-is-dr-jeffress.

9. "Show Your Support for Donald Trump," Donald J Trump for President, June 21, 2016, https://web.archive.org/web/20170118140319/https:// www.donaldjtrump.com/press-releases/trump-campaign-announces -evangelical-executive-advisory-board.

10. "Remarks by President Trump at Dinner with Evangelical Leaders," The White House, August 27, 2018, https://www.whitehouse.gov/brief ings-statements/remarks-president-trump-dinner-evangelical-leaders.

11. Jacob Pramuk, "Trump Warns North Korea Threats 'Will Be Met with Fire and Fury,'" CNBC, August 9, 2017, https://www.cnbc.com/2017/08/08 /trump-warns-north-korea-threats-will-be-met-with-fire-and-fury.html.

12. Sarah Pulliam Bailey, "'God Has Given Trump Authority to Take Out Kim Jong Un,' Evangelical Adviser Says," Washington Post, August 9, 2017, https://www.washingtonpost.com/news/acts-of-faith/wp/2017/08/08 /god-has-given-trump-authority-to-take-out-kim-jong-un-evangelical -adviser-says/?fbclid=IwAR3udKfe-EDOYm9-79FsnQE89RicqEd dhrnYRx3mSGoP1QjSApsWagIJ1gk&tid=sm_fb&utm_term=.4282a0d dc98d.

13. Andrew Silow-Carroll, "Who Is King Cyrus, and Why Did Netanyahu Compare Him to Trump?" Times of Israel, March 8, 2018, https://www .timesofisrael.com/who-is-king-cyrus-and-why-is-netanyahu-comparing -him-to-trump/.

14. Donald Trump (@realDonaldTrump). 2018. "'The Faith of Donald Trump,' a book just out by David Brody and Scott Lamb." Twitter, February 19, 2018, 6:48 p.m. https://twitter.com/realdonaldtrump/status/965780052 167286784?lang=en.

15. David Brody and Scott Lamb, The Faith of Donald J. Trump (New York: Broadside Books, 2018), 171–76.

16. Mitt Romney, "Mitt Romney: The President Shapes the Public Character of the Nation. Trump's Character Falls Short," Washington Post, Janu- ary 1, 2019, https://www.washingtonpost.com/opinions/mitt-romney-the -president-shapes-the-public-character-of-the-nation-trumps-character -falls-short/2019/01/01/37a3c8c2-0d1a-11e9-8938-5898adc28fa2_story .html?utm_term=.495e5e9ad32c.

Notes | 255

CHAPTER 3: THE OLD GOOD NEWS

1. Kathy Sawyer and Robert G. Kaiser, "The Republicans in Detroit," *Washington Post*, July 16, 1980, https://www.washingtonpost.com/archive /politics/1980/07/16/the-republicans-in-detroit/8d8ab688-c631-4eae -9d14-197cd947ae8f/?utm_term=.7e748b757531.

2. Robert P. Jones, *The End of White Christian America* (New York: Simon & Schuster, 2017), 11.

3. Daniel Williams, *God's Own Party: The Making of the Christian Right* (New York: Oxford University Press, 2010), 2.

4. Bill Clinton, "Bill Clinton Speech," CNN, August 17, 1998, http://www .cnn.com/ALLPOLITICS/1998/08/17/speech/transcript.html.

5. Thomas B. Edsall, "Resignation 'Too Easy,' Robertson Tells Christian Coalition," *Washington Post*, September 19, 1998, http://www.washington post.com/wp-srv/politics/special/clinton/stories/coalition091998.htm.

6. Hannah Rosin, "Scandal Forces Religious Issues into Public Eye," *Washington Post*, August 19, 1998, http://www.washingtonpost.com/wp-srv /politics/special/clinton/stories/ministers081998.htm.

7. James Dobson, "From Dr. James Dobson," University at Buffalo, September 1998, http://ontology.buffalo.edu/smith/clinton/character.html.

8. Ibid.

9. William Mattox Jr., "Honey, I Shrunk the Presidency," *USA Today*, August 13, 1998, p. 1A.

10. "Ten Commandments Judge Removed from Office," CNN, November 14, 2003, http://www.cnn.com/2003/LAW/11/13/moore.tencom mandments/.

11. Marc Fisher, "For Some Evangelicals, a Choice Between Moore and Morality," *Washington Post*, November 16, 2017, https://www.washingtonpost .com/politics/for-some-evangelicals-a-choice-between-moore-and-morality /2017/11/16/27a28a16-cadc-11e7-b0cf-7689a9f2d84e_story.html?utm _term=.2a04aa8064f1.

12. Jeremy Weber, "Roy Moore Was 'a Bridge Too Far' for Alabama Evangelicals," *Christianity Today*, December 13, 2017, https://www.christianity today.com/news/2017/december/roy-moore-loss-alabama-evangelicals -senate-election-mohler.html.

13. Pat Buchanan, "1992 Republican National Convention Speech," Buchanan .org, August 17, 1992, https://web.archive.org/web/20061012133633/ http://www.buchanan.org/pa-92-0817-rnc.html.

14. "Public Opinion on Abortion," Pew Research Center's Religion & Public Life Project, October 15, 2018, http://www.pewforum.org/fact-sheet /public-opinion-on-abortion.

15. "Abortion | Data and Statistics | Reproductive Health | CDC," Centers for Disease Control and Prevention, https://www.cdc.gov/reproductive health/data_stats/abortion.htm.

16. "Changing Attitudes on Gay Marriage." Pew Research Center's Religion & Public Life Project, June 26, 2017, http://www.pewforum.org/fact-sheet /changing-attitudes-on-gay-marriage.

CHAPTER 5: THE ALTAR OF WINNING

1. Scot Vorse, "The 10 Most Important Reasons Trump Would Make a Great President," Breitbart News, July 15, 2015, https://www.breitbart .com/politics/2015/07/15/the-10-most-important-reasons-trump-would -make-a-great-president/.

2. Chauncey DeVega, "The Secret History of 'Cuckservative,'" Salon, August 9, 2015, https://www.salon.com/2015/08/09/the_secret_history _of_cuckservative_the_fetish_that_became_a_right_wing_rallying_cry/.

3. David French, "The Price I've Paid for Opposing Donald Trump," National Review, October 21, 2016, https://www.nationalreview.com/2016/10 /donald-trump-alt-right-internet-abuse-never-trump-movement/.

4. Wayne Grudem, "Why Voting for Donald Trump Is a Morally Good Choice," Townhall, July 28, 2016, https://townhall.com/columnists /waynegrudem/2016/07/28/why-voting-for-donald-trump-is-a-morally -good-choice-n2199564.

5. Christianity Today editors, "James Dobson: Why I Am Voting for Donald Trump," Christianity Today, September 23, 2016, http://www.christianity today.com/ct/2016/october/james-dobson-why-i-am-voting-for-donald -trump.html.

6. Samuel Smith, "James Robison Thinks Donald Trump Was Advised Not to 'Look Like Some Weepy Christian' (interview)," Christian Post, November 4, 2016, https://www.christianpost.com/news/james-robison-donald -trump-advised-not-to-look-like-some-weepy-christian-interview-171307/.

7. Myriam Renaud, "Myths Debunked: Why Did White Evangelical Christians Vote for Trump?" University of Chicago Divinity School, January 19, 2017, https://divinity.uchicago.edu/sightings/myths-debunked-why-did -white-evangelical-christians-vote-trump.

8. Evangelicals Rally to Trump; Religious Nones 'Back' Clinton. PDF. Pew Research Center, July 13, 2016, http://assets.pewresearch.org/wp-content /uploads/sites/11/2016/07/June-Religion-and-Politics-FULL-REPORT.pdf.

CHAPTER 6: STATE OF THE CHURCH

1. Tara Isabella Burton, "Poll: White Evangelical Support for Trump Is at an All-time High," Vox, April 20, 2018, https://www.vox.com/identities /2018/4/20/17261726/poll-prri-white-evangelical-support-for-trump-is -at-an-all-time-high.

2. Robert P. Jones (@robertpjones). 2018. "White evangelicals hanging on." November 6, 2018, 7:07 p.m. https://twitter.com/robertpjones/status /1060005699877507072.

3. Ibid.

4. Bob Smietana, "Despite Mike Pence, Most Evangelical Pastors Are Not Ready to Vote Trump," Christianity Today, May 22, 2018, http://www .christianitytoday.com/news/2016/october/mike-pence-most-evangelical -pastors-undecided-trump-clinton.html.

5. Sarah Eekhoff Zylstra and Jeremy Weber, "Top 10 Stats Explaining the Evangelical Vote for Trump or Clinton," Christianity Today, May 31, 2017,

http://www.christianitytoday.com/news/2016/november/top-10-stats
-explaining-evangelical-vote-trump-clinton-2016.html.

6. "Fixed Point Foundation," Facebook, https://www.facebook.com/pg/fixed
point/about/?ref=page_internal.

7. Larry Alex Taunton, "Listening to Young Atheists: Lessons for a Stronger Christianity," *Atlantic*, May 29, 2018, https://www.theatlantic.com
/national/archive/2013/06/listening-to-young-atheists-lessons-for-a
-stronger-christianity/276584/.

8. Greg Garrison, "Founder of Christian Ministry Resigns, Admits Inappropriate Behavior." Al.com, February 6, 2018, https://www.al.com/living
/2018/02/founder_of_christian_ministry.html.

9. John Pavlovitz, "White Evangelicals, This Is Why People Are Through
with You," JohnPavlovitz.com, December 3, 2018, https://johnpavlovitz
.com/2018/01/24/white-evangelicals-people.

10. Michael Gerson, "The Trump Evangelicals Have Lost Their Gag Re-
flex," *Washington Post*, January 22, 2018, https://www.washingtonpost
.com/opinions/the-trump-evangelicals-have-lost-their-gag-reflex/2018
/01/22/761d1174-ffa8-11e7-bb03-722769454f82_story.html?utm
_term=.d927a65a8648.

11. Michelle Goldberg, "Of Course the Christian Right Supports Trump,"
New York Times, January 26, 2018, https://www.nytimes.com/2018/01
/26/opinion/trump-christian-right-values.html.

12. Patrick Kampert, "After Trump, I Can't Relate to My Evangelical Faith,"
Chicago Tribune, November 15, 2016, https://www.chicagotribune.com
/news/opinion/commentary/ct-trump-evangelical-christians-faith-chal
lenged-perspec-1116-md-20161115-story.html.

13. Michelle Goldberg, "Of Course the Christian Right Supports Trump."

14. Phil Zuckerman, "The Immorality of Evangelical Christians in the Age of
Trump," *Huffington Post*, February 21, 2018, https://www.huffingtonpost
.com/phil-zuckerman/the-immorality-of-evangel_b_14658334.html.

15. Emily Lund, "Trump Won. Here's How 20 Evangelical Leaders Feel,"
Christianity Today, February 28, 2019, https://www.christianitytoday
.com/ct/2016/november-web-only/trump-won-how-evangelical-leaders
-feel.html.

16. The Republican excitement over job numbers curiously excludes certain
metrics such as worker participation, which conservatives railed on until
somewhere around January 20, 2017.

17. Ben Howe, "About Last Night . . ." *RedState*. November 9, 2016, https://
www.redstate.com/aglanon/2016/11/09/last-night . . . /.

18. Quinnipiac University, "QU Poll Release Detail," QU Poll, January 25,
2018, https://poll.qu.edu/national/release-detail?ReleaseID=2516.

CHAPTER 7: TRUE VICTORY

1. Dennis McCallum, "Geisler's Three Schools of Principlized Ethics," Xe-
nos Christian Fellowship, https://www.xenos.org/essays/geislers-three
-schools-principlized-ethics.

2. Katie Reilly, "You May Have More Money in Your Paycheck Due to Tax Cuts | Money," *Time*, February 2, 2018, http://time.com/money/5130699 /new-tax-bill-paycheck-irs-calculator/.

3. Jennifer Steinhauer, "Bill Clinton Should Have Resigned over Lewinsky Affair, Kirsten Gillibrand Says," *New York Times*, November 17, 2017, https://www.nytimes.com/2017/11/16/us/politics/gillibrand-bill-clinton -sexual-misconduct.html.

4. Louis C. K., "Live at the Comedy Store (2015)—Transcript," May 4, 2017, http://scrapsfromtheloft.com/2017/05/04/louis-c-k-live-at-the-comedy -store-2015-transcript.

5. Melena Ryzik, Cara Buckley, and Jodi Kantor. "Louis C. K. Is Accused by 5 Women of Sexual Misconduct," *New York Times*, November 9, 2017, https://www.nytimes.com/2017/11/09/arts/television/louis-ck-sexual -misconduct.html?_r=0.

AFTERWORD: LOVE YOUR ENEMIES

1. C. S. Lewis, *Mere Christianity* (New York: HarperOne, 2018), 8.

INDEX

ABOUT THE AUTHOR

Ben Howe is a writer, podcaster, and filmmaker, as well as the founder of Howe Creative, a video production company. He has appeared as a commentator on CNN, MSNBC, and Fox News, and has written guest columns for the *Atlantic*, the *Washington Examiner*, and the *Daily Beast*, among other publications. You can follow him on Twitter at @BenHowe.